The Basic Book of
Power Mechanics

Jay Webster

American Technical Publishers, Inc.
Alsip, Illinois 60658

PRINTED IN THE UNITED STATES OF AMERICA

CONTENTS

4   96   Cooling and Lubrication

THE BASIC BOOK OF POWER MECHANICS is part of an integrated series of Industrial Arts textbooks designed to teach basic skills to beginning students. Its major objectives are: career exploration, development of consumer awareness, manipulative skills, and craftsmanship. The philosophy of THE BASIC BOOK OF POWER MECHANICS is based on a recent, nationwide survey in which power mechanics teachers at all levels were asked to outline the courses they taught and let us know what types of instructional materials they actually needed. The result is a highly-visual text with a controlled reading level that will help insure student success.

The author and the publisher wish to acknowledge and thank the following corporations for their assistance and cooperation: Altec-Lansing, Benson Aircraft, BMW, Briggs and Stratton, Brodhead-Garrett, Champion Spark Plug Company, Chrysler Corporation, Clinton, Columbia, Outboard Marine Corp., Fiat Motors, Heald, Jacobsen, McCulloch, Oldsmobile Division, Onan, Premier, Rotorway, Suzuki, Toro, and Volkswagen of America.

The Publisher

INTRODUCTION 1

What's your idea of a sharp automobile? A Corvette with a turbocharger? A VW with a diesel? A full race Lotus? Perhaps a '32 Ford chassis with four on the floor behind a big block Chevrolet V-eight?

You may think something else is a better machine. All cars have one thing in common: they depend on their engines. Everything you want to know about engines is included in power mechanics.

Maybe you'd like to build your own helicopter, or pep up the family ski boat. Well, that's power mechanics too.

Of course, it's easier to start by fixing the family power lawn mower. It's only a small step from lawn mowers to motorcycle engines and another step to car and truck and airplane power mechanics.

You can use your knowledge of power mechanics for emergency repairs on the side of the road, or become a professional mechanic. If you go to college to study power mechanics you may become an automobile engineer or an aircraft engineer. There are thousands of careers (lifetime jobs) for men and women who study power mechanics.

Do you know what a family car and a power lawn mower have in common? They both have an engine that uses fuel to make power (figure 1-1). They are both power machines. Power mechanics is the study of machines that make power.

The study of power mechanics is the study of airplanes, boats, cars, motorcycles, go-carts, mopeds, and snowmobiles. Learning how any one of these machines works will help you to understand how all power-driven machines work (figure 1-2).

This book explains small engines. Big engines work the same way as small ones, they just have more parts.

There are four basic things you need to know in order to understand all engines:

- How they work
- The names of their parts
- How they get spark and fuel
- How they are cooled and lubricated

Learning about small engines will help you find out why any engine won't run—or runs badly. You will learn how to make carburetor adjustments, change oil, clean air filters, gap spark plugs, and do many other jobs that are common to most engines.

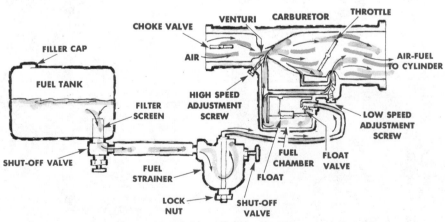

Figure 1-1: All power machines have a system to deliver fuel.

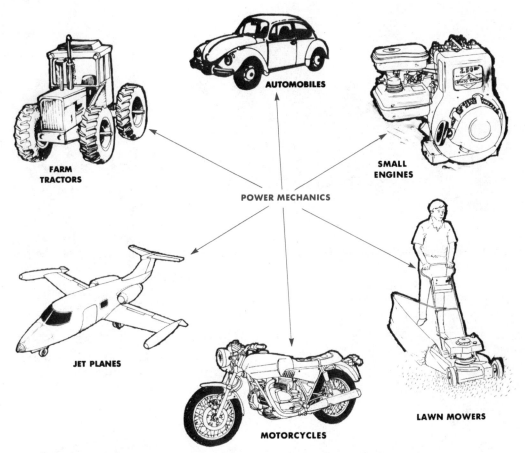

FARM
TRACTORS

AUTOMOBILES

SMALL
ENGINES

POWER MECHANICS

JET PLANES

MOTORCYCLES

LAWN MOWERS

Figure 1-2: The study of power mechanics is the study of engines—large and small.

Figure 1-3: Since all power machines are similar, the study of small engines helps you to understand how big ones work.

Figure 1-4: The lawnmower can be a simple power machine or a complicated piece of heavy duty machinery.

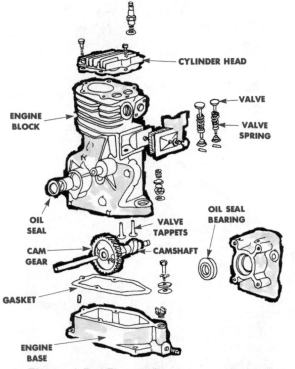

Figure 1-5: The study of power machines is a study of their parts.

Everything you learn to do on a small engine is also done on big power machines. Everything you learn about

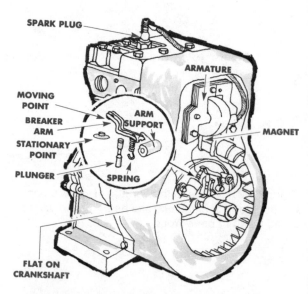

Figure 1-6: Many power machines have a system to provide an electrical spark.

small engines will help you understand all engines. Big engines have the same problems as small ones, and they are repaired the same way.

1. What do lawn mowers and cars have in common?
2. Name the four things you need to know to understand all engines.
3. What is power mechanics?

4. Why should you study power mechanics?

██ **Activities for Unit 1**

1. Make a list of all the power driven machines you can think of that use fuel to make power.
2. Make a list of all the power driven machines your family owns.
3. Make a list of all the things you would like to learn how to do on an engine.

Power mechanics is not dangerous as long as you always think—"SAFETY FIRST!" Safety First means preventing accidents before they happen by always asking yourself, "Is what I am doing safe?" If you don't know, ask your teacher (figure 2-1).

The misuse of ordinary <u>hand tools</u> causes a lot of accidents. Greasy tools are easy to drop. A tool that falls into a moving engine may shoot out like a rocket and injure someone. Sharp tools are as dangerous as sharp knives. Treat tools right and they'll treat you right. Figure 2-2 is a list of basic safety tips. The better you know them, the safer you will be.

<u>Compressed air</u> is a useful tool in power mechanics shops, but it can also be dangerous if it is used carelessly. Figure 2-3 is a guide to the proper, safe use of compressed air. Pay attention to it!

SAFETY GLASSES

FACE SHIELD

Figure 2-1: Safety glasses or a face shield should be used for eye protection when cutting, filing, using compressed air, or running an engine.

- BE SURE YOUR HANDS ARE CLEAN OF DIRT, GREASE AND OIL.

- USE THE PROPER TYPE AND SIZE HAND TOOL.

- MAKE SURE THE TOOLS YOU ARE GOING TO USE ARE SHARP AND IN GOOD CONDITION.

- USE SHARP-EDGED OR POINTED TOOLS WITH CARE.

- WHEN USING A SHARP-EDGED TOOL MAKE SURE TO POINT THE EDGE AWAY FROM YOURSELF AND YOUR CLASSMATES.

- WEAR A FACE SHIELD OR SAFETY GLASSES WHEN FILING OR CUTTING METAL. ARRANGE YOUR WORK SO THAT YOUR CLASSMATES ARE PROTECTED FROM FLYING CHIPS.

- PASS TOOLS TO CLASSMATES WITH THE HANDLES FIRST.

- CLAMP SMALL WORK ON A BENCH OR SECURE IT IN A VISE.

Figure 2-2: Safety rules for using hand tools.

- CHECK ALL HOSE CONNECTIONS BEFORE TURNING ON THE AIR.

- HOLD THE AIR HOSE NOZZLE TO PREVENT IT FROM SLIPPING WHILE TURNING AIR ON OR OFF.

- DO NOT LAY THE HOSE DOWN WHILE THERE IS PRESSURE IN IT. IT MIGHT WHIP ABOUT AND STRIKE SOMEONE.

- DO NOT USE AIR TO DUST OFF HAIR OR CLOTHING, OR TO SWEEP THE FLOOR.

- WEAR SAFETY GLASSES WHEN USING COMPRESSED AIR.

Figure 2-3: Safety rules for using compressed air.

- CHECK WITH YOUR INSTRUCTOR BEFORE STARTING AN ENGINE.

- CHECK THE FUEL LINE FOR POSSIBLE LEAKS.

- EXHAUST ALL GASES TO THE OUTSIDE OF THE BUILDING, AND BE SURE THERE IS ADEQUATE VENTILATION WHEN YOU RUN AN ENGINE.

- KEEP YOUR HEAD AND HANDS AWAY FROM MOVING PARTS.

- DO NOT RUN AN ENGINE AT HIGH SPEED FOR A LONG TIME.

- WEAR FACE AND EAR PROTECTION WHEN RUNNING AN ENGINE AT HIGH SPEED.

Figure 2-4: Safety rules for running an engine.

Running an engine is what power mechanics is about. The safe way to run any engine in the class room is outlined in figure 2-4. These rules protect against:

- Exploding an engine by running it too fast

- Injuries from broken engine parts

- Carbon monoxide poisoning from the engine exhaust

Carbon monoxide is a colorless, odorless gas that can cause suffocation. It comes from the engine exhaust. Carbon monoxide poisoning is prevented by making sure exhaust gases can escape from, and fresh air can get into, the room.

Fire prevention and fire fighting methods are important to know. Fuels that run engines burn easily and are a hazard in every power mechanics shop.

Figure 2-5: Too much noise can be a hazard. Ear protectors cut the noise level.

To burn, fires need oxygen, fuel, and temperature. To put out a fire, you must remove one of these elements. Most fire extinguishers put out fires by preventing oxygen from reaching the flames. Every shop should have a fire extinguisher. Before you start working in the shop you should know where the fire extinguishers (figures 2-6) are, and how to use them. The rules for fire prevention are shown in figure 2-7.

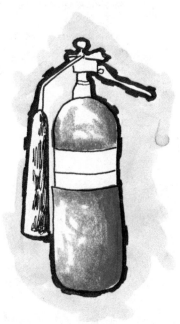

Figure 2-6: Know where the fire extinguisher is and how to use it.

- STORE ALL COMBUSTABLE LIQUIDS IN A FIREPROOF ROOM OR CABINET.
- STORE GASOLINE IN A SAFETY CAN (ONE WITH A TOP WHICH IS HELD CLOSED WITH A SPRING).
- PLACE RAGS WITH OIL, GASOLINE, PAINT, OR SOLVENTS ON THEM INTO A SAFETY CONTAINER.
- KEEP THE TOP OF OIL CANS CLEAN.

Figure 2-7: Safety rules for fire prevention.

SELF CHECK

1. How should you handle sharp tools?
2. What is the danger of running an engine too fast?
3. Why is carbon monoxide unsafe?
4. List three things a fire must have.

Activities for Unit 2

1. Draw a sketch of your school shop showing where the fire extinguisher and other safety equipment can be found.
2. List the steps you would follow to start an engine safely in your school shop.
3. Make a list of the safety rules you think are most important.

Figure 2-8: Putting fire hazardous material in the proper container prevents fires.

An engine is a machine that changes heat into mechanical power. There are several types of engines. The main engine in use today, and the one this book is about, is the internal combustion engine (figure 3-1).

When combustion, the burning of air and fuel, takes place inside an engine (internally), the engine is called an internal combustion engine (figure 3-2). The modern engine has five main working parts:

Figure 3-1: Engines change heat into mechanical energy.

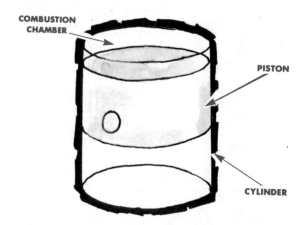

Figure 3-2: The space where fuel and air is burned in an engine is the combustion chamber.

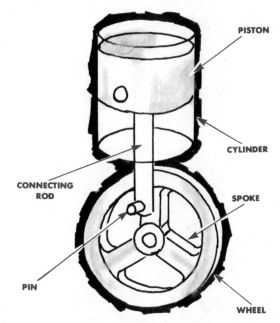

Figure 3-3: The connecting rod attaches the piston to a spoke on a wheel.

- The cylinder
- The piston
- The connecting rod
- The crankshaft
- The flywheel

The cylinder and piston become an engine when the piston is connected to something that uses the power of combustion for work. The piston is usually connected to a rod called a connecting rod. The connecting rod is attached to a pin on the spoke of a wheel (figure 3-3). Combustion forces the piston down and turns the wheel (figure 3-4).

In a modern engine the rod connects to an offset shaft called a crankshaft (figure 3-5). A crankshaft works the same as a wheel. It turns the downward motion of the piston into rotary motion. More than one rod and piston

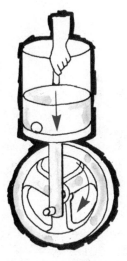

Figure 3-4: When the piston is forced down, it turns the wheel.

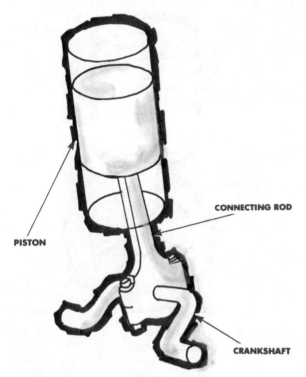

Figure 3-5: The connecting rod is connected to an offset shaft called the crankshaft.

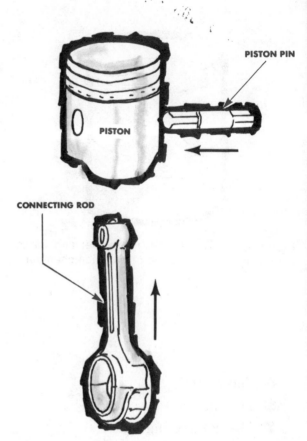

Figure 3-6: The piston pin holds the rod to the piston.

can be connected to the same crankshaft. The rod is connected to the piston by a <u>piston pin</u> (figure 3-6).

When you ride a bicycle, your leg acts like the piston and rod. The pedals are like the special bends in the crankshaft (figure 3-7).

There's only one problem with our basic engine. The piston will go down the cylinder, but it won't come back up again. We want our engine to work more than one time, and this means we must bring the piston back to the top of the cylinder. We do this with a <u>flywheel</u>. The flywheel is a heavy wheel mounted on the end of the crankshaft as shown in figure 3-8. The flywheel turns with the crankshaft. Because it is heavy it does not stop turning when the piston reaches the bottom of the cylinder. This weight causes the piston to go back up to the top of the cylinder.

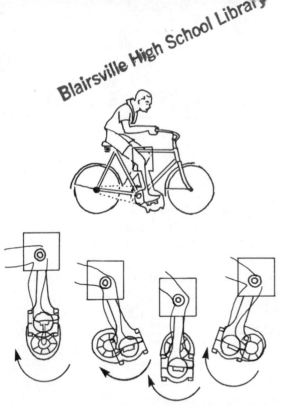

Figure 3-7: When you pedal a bike your knee acts like a piston pin, your leg acts like a rod, and the pedals work the same as the crankshaft.

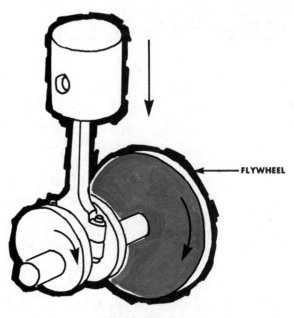

Figure 3-8: The heavy flywheel turns with the crankshaft. Its momentum then forces the piston back up the cylinder.

SELF CHECK

1. List the five main parts of an engine.
2. Where is the combustion chamber?
3. What is a crankshaft?
4. What does the flywheel do?

Activities for Unit 3

1. Make a sketch of a basic engine and point out the five main working parts.

2. Using a cut-away engine model, identify the five main working parts of an engine.

3. Look through some of the small engine parts in your school shop and point out the five working parts of an engine.

A stroke in any engine is the movement of a piston from the top of the cylinder to the bottom (figure 4-1). Movement of the piston from the bottom to the top of the cylinder is also a stroke (figure 4-2).

The four-stroke cycle engine is one in which the piston moves from the top to the bottom of the cylinder (stroke one), rises back to the top (stroke two), returns to the bottom (stroke three), and rises to the top again (stroke four) during one complete engine cycle.

A cycle is a series of actions that are repeated over and over.

For our basic engine to work, it must have two holes in the top of the cylinder. We must be able to open and close these holes as needed. One hole will be used to let air and fuel into the cylinder. It is called the intake valve port or passage. The other hole is used to let burned air and fuel out of the cylinder. It is called the exhaust valve port or passage.

The intake stroke starts with the pis-

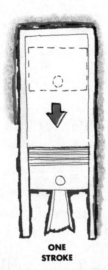

Figure 4-1: When a piston moves from the top of the cylinder to the bottom it has completed one stroke.

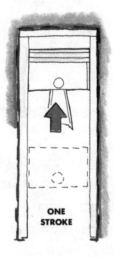

Figure 4-2: Movement of the piston from the bottom of the cylinder to the top is also one complete stroke.

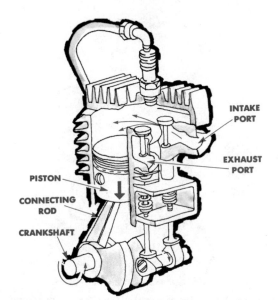

INTAKE
PORT

EXHAUST
PORT

PISTON

CONNECTING
ROD

CRANKSHAFT

Figure 4-3: On the intake stroke, the piston moves down, drawing air and fuel into the cylinder through the intake valve port. The crankshaft rotates half a turn.

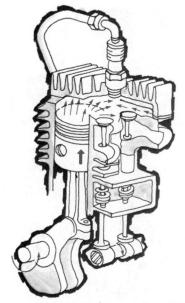

Figure 4-4: The compression stroke squeezes the fuel and air into the small space of the combustion chamber. Both ports are closed and the crankshaft has rotated a full turn.

ton at the top of the cylinder. The piston begins to move down the cylinder, the intake valve port opens, and the piston moving down the cylinder creates a reduced pressure that draws air and fuel directly into the cylinder (figure 4-3).

When the piston gets to the bottom of its stroke, the cylinder above is full of air and fuel. The crankshaft has made half a turn, and the intake valve port closes.

The next stroke is the compression stroke. As the piston begins to move back up the cylinder, both valve ports are closed, and it squeezes, or compresses, the air and fuel (figure 4-4).

The tighter the air and fuel mixture is squeezed, the hotter it will burn.

At the end of the compression stroke the piston is at the top of the cylinder again. Both valve ports are still closed, and the air and fuel are compressed in the combustion chamber. The crankshaft has completed one full turn.

The power stroke starts at the end of the compression stroke. The air-fuel mixture is set on fire, and the burning gases force the piston back down the cylinder (figure 4-5). The intake and exhaust ports remain closed during the power stroke.

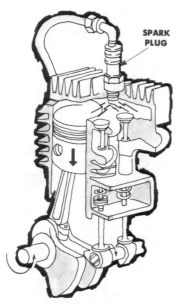

SPARK PLUG

Figure 4-5: The power stroke begins when the fuel and air are ignited by a spark plug. The hot gases expand, and push the piston down. Both valve ports are closed, and the crankshaft makes a half turn.

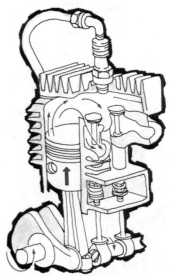

Figure 4-6: On the exhaust stroke, the piston forces the burned gases out the exhaust valve port which is now open. The crankshaft has rotated a full turn.

The exhaust stroke is the fourth stroke in the four-stroke cycle. The piston starts back up the cylinder, and the exhaust valve port opens letting the rising piston push the hot gases out of the cylinder (figure 4-6). As the piston reaches the top of the cylinder the exhaust valve port closes. The crankshaft has now completed two full turns and four full strokes. The next stroke is an intake stroke, it begins the cycle over again.

Remember that the four-stroke cycle engine requires all four strokes to develop power, and that these four strokes are tied to the intake and exhaust valve ports. All four-stroke cycle engines develop power in the following order (figure 4-7):

- Intake stroke: Intake valve port open and the piston is moving down
- Compression stroke: Both valve ports are closed and the piston is moving up
- Power stroke: Both valve ports are closed and the piston is being forced down by the burning gases
- Exhaust stroke: Exhaust valve port open and the piston is moving up

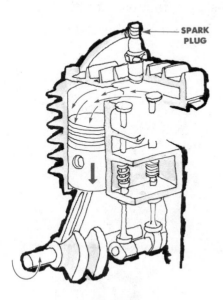

SPARK
PLUG

INTAKE STROKE

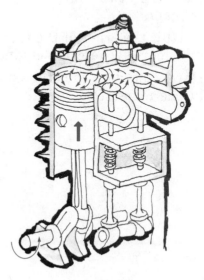

COMPRESSION STROKE

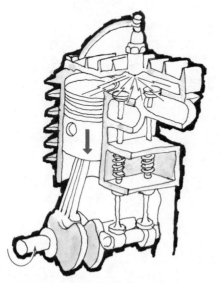

POWER STROKE

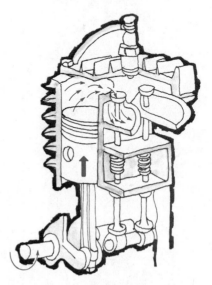

EXHAUST STROKE

Figure 4-7: Here are the four strokes of a four-stroke engine cycle.

SELF CHECK

1. What is a stroke?
2. List the two valve ports in a cylinder.
3. What is the order of the four strokes in a four-stroke cycle engine?
4. How often does the crankshaft turn in one four-stroke cycle?

Activities for Unit 4

1. Make a sketch of each of the four strokes of a four-stroke cycle engine.
2. Write an explanation for each stroke.

Not all engines use the four-stroke cycle principle. Many develop power with just two piston strokes. These engines are called two-stroke cycle engines.

The basic parts of a two-stroke cycle engine are the same as those of a four-stroke cycle engine. These parts are the cylinder, piston, connecting rod, crankshaft, and flywheel. As with the four-stroke cycle engine the piston and cylinder form a combustion chamber (figure 5-1).

As you can guess from its name, the crankcase is a box that encloses the crankshaft. In a two-stroke cycle engine the crankcase is very important because the air and fuel mixture is compressed there before it passes into the cylinder through the intake port. The intake port is connected to the crankcase by a passageway. This passageway is called a <u>by-pass or transfer port</u>.

When the piston moves up in the cylinder it draws air and fuel into the crankcase through the intake port. As shown in figure 5-2, some pistons have a contoured crown. This contoured crown is called a deflector fin. It helps the air-fuel mixture flow in and out of the combustion chamber. The sharp angle on the intake side of the deflector fin forces the fresh air-fuel mixture towards the top of the combustion

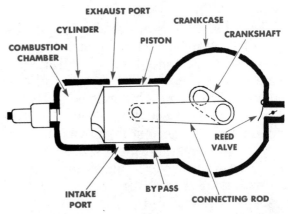

Figure 5-1: The basic parts of the two-stroke engine are the same as the four-stroke. The main differences are the reed valve on the crankcase and the bypass from the crankcase to the intake port.

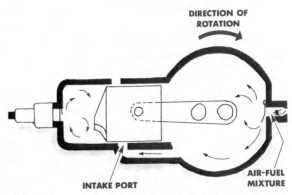

Figure 5-2: The rising piston compresses fuel and air in the cylinder. At the same time, it causes a vacuum in the crankcase, pulling in air and fuel through the reed valve.

chamber. The gentle angle of the contour on the exhaust side allows the burnt mixture to flow out easily. The rising piston covers the intake and exhaust ports. Air and fuel above the piston is trapped in the combustion chamber, and the rising piston compresses it (figure 5-3).

The fuel is ignited at the top of the piston's stroke. The burning gases force the piston down turning the crankshaft (figures 5-4). As the piston moves down the cylinder it compresses the air and fuel which is drawn into the crankcase as the piston moves up.

As the piston moves down the cylinder it uncovers the exhaust port, allowing the burnt air-fuel mixture to escape (figure 5-5). Note how the deflector fin controls the flow of the air-fuel mixture and exhaust gases. The piston continues its stroke towards the bottom of the cylinder and uncovers the intake port. Now comes the tricky part. The air-fuel mixture, which is under pressure in the crankcase, flows through the passageway, through the intake port, and into the cylinder. The fresh air-fuel mixture pushes the burned air-fuel mixture out the exhaust port. The rising piston compresses it, and starts the cycle again.

The valve in the crankcase of a two-stroke cycle engine is called a <u>reed valve</u>. It is made of thin metal, and is not found in the four-stroke cycle engine. Its purpose is to let air and fuel into the two-stroke cycle crankcase. Figures 5-6 and 5-7 show how a reed valve works.

When the piston rises, a vacuum is created in the crankcase, the reed valve

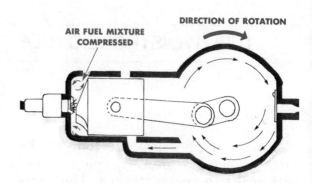

Figure 5-3: At the top of the stroke the air-fuel mixture is squeezed into the combustion chamber.

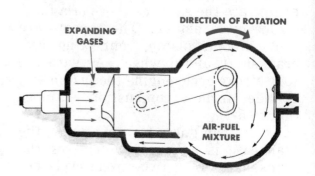

Figure 5-4: The burning air-fuel mixture forces the piston down, turning the crankshaft and compressing the air-fuel mixture in the crankcase. Note how the piston fin directs the flow of the air-fuel mixture and the exhaust gases.

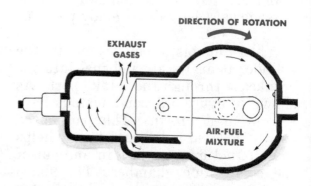

Figure 5-5: At the bottom of the power stroke the exhaust gases escape from the cylinder while the air-fuel mixture is forced in.

is pushed open by the outside air pressure, and the air-fuel mixture flows into the crankcase. This happens on the engine's compression stroke (figure 5-6).

The reed valve closes when the piston is forced downward by the burning air-fuel mixture (figure 5-7). This occurs on the power stroke. Figure 5-8 shows a few of the different types of reed valves.

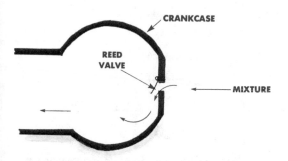

Figure 5-6: On the compression stroke vacuum is created in the crankcase. This vacuum opens the reed valves allowing fuel and air to enter.

Some two-stroke cycle engines have holes or slots in the skirt area of their pistons. This type of engine is called a loop scavenge engine. A loop scavenge engine works just like any other two-stroke cycle engine. The only difference is that the air-fuel mixture travels (or loops) through the holes in the pistons.

The operation of a loop scavenged engine is shown in figure 5-9. When the piston is at the bottom of its stroke, the holes in the piston skirt line up with passages in the cylinder. The air-fuel mixture that was slightly compressed during the power stroke of the piston, passes into the cylinder passages through the holes in the piston, and into the cylinder. By introducing the air-fuel mixture at two sides of the cylinder, the need for a piston with a deflector fin is eliminated. The loop scavenge type of engine produces a greater horsepower per unit of weight by providing a more complete removal of the exhaust gases at the end of the power stroke.

Figure 5-7: The power stroke of the piston compresses the air-fuel mixture in the crankcase. This pressure closes the reed valve.

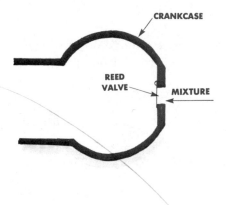

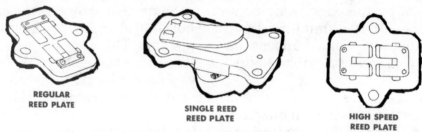

REGULAR REED PLATE

SINGLE REED REED PLATE

HIGH SPEED REED PLATE

Figure 5-8: Although they may look different, all reed valves work in the same way.

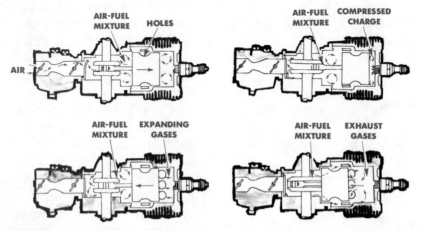

AIR-FUEL MIXTURE   HOLES

AIR

AIR-FUEL MIXTURE   COMPRESSED CHARGE

AIR-FUEL MIXTURE   EXPANDING GASES

AIR-FUEL MIXTURE   EXHAUST GASES

Figure 5-9: In this engine, the air-fuel mixture passes into the cylinder through holes in the piston. It works the same way as the bypass.

SELF CHECK

1. How many strokes are used to develop power in a two-stroke engine?

2. How does the crankcase of a two-stroke cycle engine differ from the crankcase of a four-stroke cycle engine?

3. What does the by-pass or transfer port do?

4. How does a reed valve work?

Activities for Unit 5

1. Make a sketch of the two strokes in a two-stroke cycle engine.

2. Write an explanation for each of the strokes.

3. Using a cut-away engine model, turn the crankshaft and identify the beginning and end of each of the strokes.

Automotive engines have more parts than the basic engine we have studied so far. Very few of these parts can be seen from the outside of the engine. They must be shown by cut-away and exploded pictures.

A cut-away view of a two-stroke cycle engine is shown in figure 6-1. A four-stroke cycle engine is shown in figure 6-2.

An exploded view (figure 6-3) shows the parts floating in air in the position in which they are to be assembled.

The main parts of a small engine are:

Figure 6-1: The cut-away view of a two-stroke cycle engine shows most of its working parts.

Figure 6-2: Cut-away views make it possible to see how different parts work with each other. This is a four-stroke cycle engine.

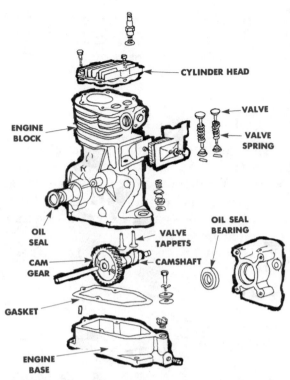

Figure 6-3: This exploded view of a four-stroke cycle engine shows how the parts go together.

- Crankcase
- Crankshaft
- Bearings
- Cylinder
- Cylinder Assembly (block)
- Cylinder Head
- Piston
- Piston Rings
- Connecting Rods
- Piston Pin

The crankcase is the metal box or housing that holds the crankshaft (figure 6-4). There are two holes called

the main bearings in the sides of the crankcase. Some small engines have bearings or bushings inserted in these holes. When this is done, the holes are called main bearing bores, and the bearings or bushings are called the main bearings. One of the main bearings is in a side cover or plate which can be removed from the crankcase (figure 6-6). When this plate is taken off the crankshaft can be removed.

Many small engines do not have main bearings. The crankshaft rotates directly in the aluminum crankcase.

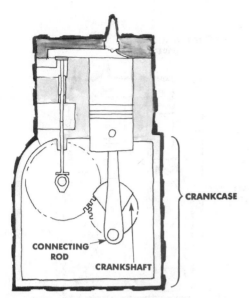

Figure 6-4: The crankcase holds the crankshaft, protecting it from damage.

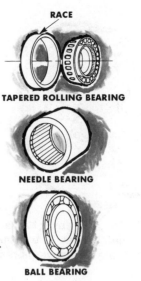

Figure 6-5: Types of main bearings.

Some engines use tapered, roller, ball, or needle bearings to support the crankshaft (figure 6-5). The more expensive small engines often use a ball bearing on the power take-off end of the crankshaft. This bearing works very hard when the engine works hard. Two-stroke cycle engines often use needle bearings for both main bearings. Larger stationary engines that turn slowly often use a tapered roller bearing.

The cylinder is the hollow tube in which the piston slides up and down. The inside surfaces of the cylinder are called the cylinder walls. The piston must fit snugly inside the cylinder, but is must be loose enough to move up and down freely. Many small engine cylinders are made of aluminum. It is light, and it transfers heat quickly. It is also very soft. Sometimes a thin, cast-iron liner called a sleeve, is used on the inside of an aluminum cylinder to keep it from wearing out too quickly. Many motorcycle engines have removable cylinders. A damaged or worn out cylinder can be easily replaced if it is removable.

A cylinder assembly (figure 6-7) is a one-piece cylinder and crankcase. The cylinder assembly is commonly called a "block".

The cylinder head (figure 6-8) is attached to the top of the cylinder with bolts (figure 6-9). Remember that the cylinder is part of the engine block. The head gasket fits between the cylinder head and the cylinder (figure

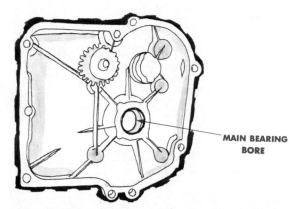

Figure 6-6: The side cover has a main bearing for the crankshaft journal. The crankshaft can be taken out after the side cover is removed.

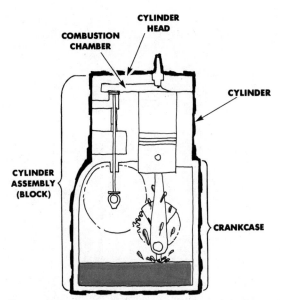

Figure 6-8: The cylinder head fits on top of the block.

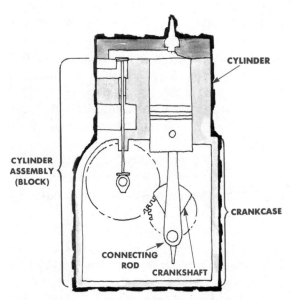

Figure 6-7: A one-piece cylinder and crankcase are called a cylinder assembly or a block.

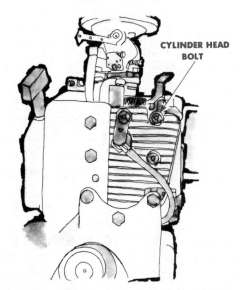

Figure 6-9: The cylinder head is held to the block with bolts.

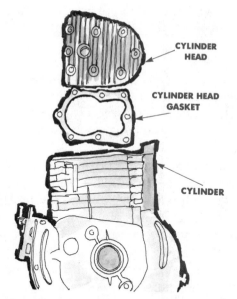

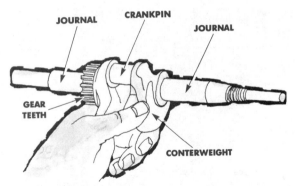

Figure 6-11: The parts of a crankshaft include the journals, and at least one crankpin and counterweight. Gear teeth are used to drive a camshaft.

Figure 6-10: The cylinder head gasket makes an airtight seal between the cylinder head and the cylinder.

6-10). It keeps the high pressure created by the burning gases from escaping between the cylinder and the cylinder head.

The crankshaft (figure 6-11) fits in the crankcase. It changes the up and down movement of the piston to rotating (round and round) movement. The ends of the crankshaft that fit into the main bearings are called journals. They must fit tightly so the oil in the crankcase will not leak out.

The crankpin is where the crankshaft connects to the connecting rod. It is the offset part of the crankshaft. The crankpin makes it possible to change the up and down movement of the piston into the rotary movement on the crankshaft. Figures 4-3 through 4-6 show this action in detail. Counterweights on the crankshaft balance the weight of the piston and connecting rod. This helps the engine run smoothly.

In some engines the crankshaft is horizontal or parallel with the ground. Other engines have a vertical (up and down) crankshaft (figure 6-12).

Pistons should be light. Most small engine pistons are made of aluminum. Figure 6-13 shows the parts of a piston. The top surface of a piston is called the piston crown. It must be able to withstand the heat and pressure of the burning air-fuel mixture. The side of the piston is called the piston skirt. The final piston part is the pin hole. The piston pin fits into the pin hole and the connecting rod is attached to it.

Piston rings are the round metal rings that fit into grooves in the side of

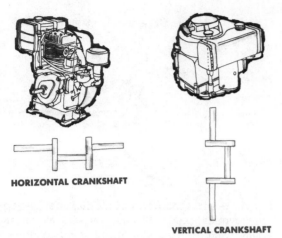

HORIZONTAL CRANKSHAFT

VERTICAL CRANKSHAFT

Figure 6-12: Crankshafts may be mounted vertically or horizontally.

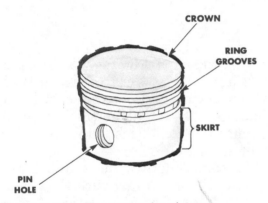

Figure 6-13: The parts of a piston.

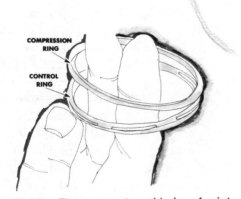

Figure 6-14: There are two kinds of piston rings: compression rings and oil-control rings.

the piston, and press against the cylinder walls to make a tight fit. The top rings are called compression rings (figure 6-14 top). They keep the burning air-fuel mixture from leaking into the crankcase. The bottom rings are called oil-control rings (figure 6-14 bottom). They have either holes or slots in them.

In a running engine, oil from the crankcase is splashed onto the cylinder walls. The oil-control rings wipe the oil from the cylinder walls to keep it out of the combustion chamber. Oil is needed to lubricate cylinder walls, but if it gets into the combustion chamber it will burn and cause smoke.

The connecting rod (figure 6-15) con-

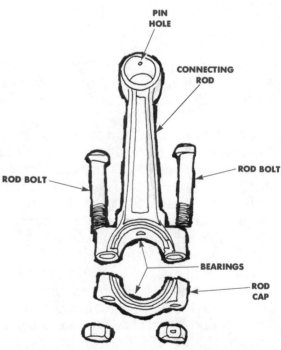

Figure 6-15: The parts of a connecting rod.

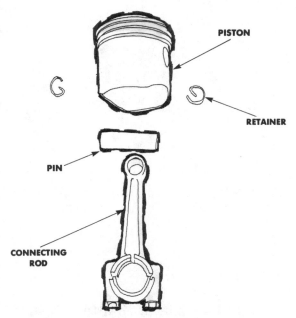

Figure 6-16: The connecting rod is attached to the piston with a pin and retaining rings.

(figure 6-16) connects the connecting rod and the piston together. It allows the rod to move back and forth as the crankshaft turns. Retaining rings hold the piston pin in place. The other end of the rod fits around the crankshaft. It is held in place by the rod cap (figure 6-15). The rod cap is held in place by connecting rod bolts.

The rod and rod cap must fit together tightly because the turning of the crankshaft creates friction. The snug fit of the rod and rod cap and the bearings between the crankshaft and the connecting rod, which most engines have, prevent friction. A cut-away picture of a complete piston-crankshaft assembly is shown in figure 6-17.

Many small engines do not have a separate connecting rod bearing. The aluminum connecting rod saddle bore is used as the bearing surface.

Figure 6-17: This view shows the complete hook-up between the piston and the crankshaft.

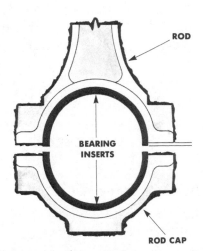

Figure 6-18: An insert bearing is used in some connecting rods.

nects the piston to the crankshaft. Like the piston, the connecting rod has a pin hole for the piston pin. The piston pin, also called the wrist pin,

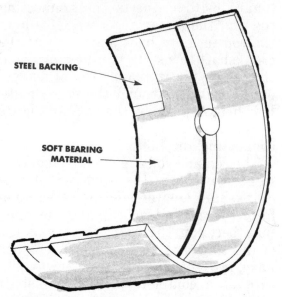

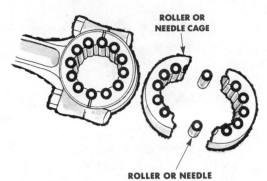

ROLLER OR
NEEDLE CAGE

STEEL BACKING

SOFT BEARING
MATERIAL

ROLLER OR NEEDLE

Figure 6-20: Needle bearings are used in the connecting rods of some two-stroke cycle engines.

Figure 6-19: An insert bearing has a steel back. It is faced with soft bearing material.

Some small engines use a precision insert bearing between the crankshaft and the connecting rod bore. The insert bearings are made in two pieces so they may be assembled around the crankshaft (figure 6-18). They consist of a steel backing with a thin layer of soft bearing material attached to the backing (figure 6-19).

Many two stroke engines use needle bearings in the saddle bore (figure 6-20). The needle bearings may be free to touch each other or they may be held apart by a cage.

Activities for Unit 6

1. Using a disassembled shop engine, identify the crankcase, block assembly, and cylinder head.

2. Using a disassembled shop engine, identify the piston, connecting rod, and crankshaft.

3. Find a compression ring and an oil-control ring. Explain what each does in an engine.

SELF CHECK

1. What is a crankcase?
2. What is a one-piece cylinder and crankcase called?
3. Where does the head gasket go?
4. List the two types of piston rings.

For an engine to run, fuel and air must be delivered to, burnt in, and removed from, the cylinder at just the right time. This is the job of the valves. In our basic engine, the valves have been described as simple holes (or ports) in the cylinder. The ports in a real engine are opened and closed by valves that work somewhat like the plug in a bathtub (figures 7-1 and 7-2).

An engine valve is a round metal plug connected to a stem (figure 7-3). The head of the valve has a tapered face that fits tightly into a valve seat. The valve seat (figure 7-4) is a tapered hole in the cylinder block. The taper of the valve matches the taper of the valve seat so a tight seal can be made.

The purpose of the valves is to let fuel and air into the cylinder and to let the burned gases out. The intake valve lets fuel and air into the cylinder, and the exhaust valve lets it out.

A camshaft is used to open and close the valves. The camshaft (figure 7-5) is a round piece of metal with bumps called cam lobes on it.

The camshaft is located in the crankcase under the valves. As the camshaft turns, the lobes raise and lower the valves (figure 7-6). The camshaft for a single cylinder engine has two lobes. One lobe works the intake valve. The other lobe works the exhaust valve.

A gear on the crankshaft drives a gear on the camshaft (figure 7-7). The

Figure 7-1: Just as a plug in a bathtub drain stops water from escaping, an engine valve keeps gases in the cylinder.

Figure 7-2: In the open position, a plug lets water out of a bathtub. In an engine the valve lets exhaust gases out and air and fuel in.

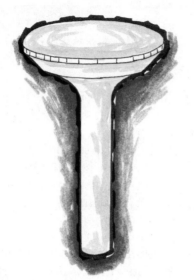

Figure 7-3: An engine valve resembles a metal bathtub plug mounted on a stem.

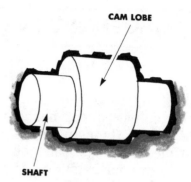

Figure 7-5: A camshaft is a metal shaft with a bump on it. The bump is called a cam or a cam lobe.

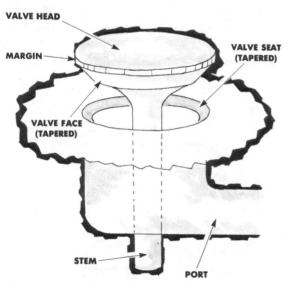

Figure 7-4: The valve seat is tapered to fit the valve face.

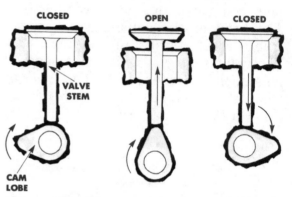

Figure 7-6: The cam lobe rotates under the valve and pushes it open.

gear on the crankshaft is of a different size than the gear on the camshaft because the four-stroke cycle engine crankshaft must make two full turns to complete the four-stroke-cycle, while the camshaft makes only one turn. Figure 7-8 shows how the gears are marked so they can be meshed together correctly.

Valve lifters (figure 7-9) are used in most engines. They are metal rods that act as a connection between the cam lobes and valve stems.

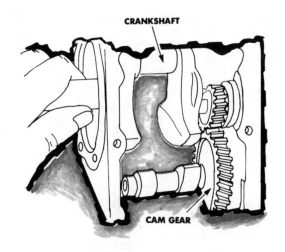

Figure 7-7: A gear on the crankshaft drives the camshaft at one-half the speed of the crankshaft. These gears are called the timing gears.

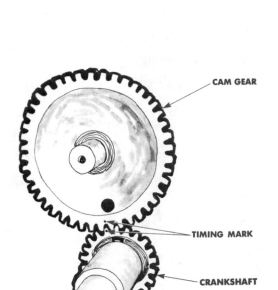

Figure 7-8: The marks on the timing gears are called timing marks. They must be lined up to make sure the camshaft opens the valves at the right time.

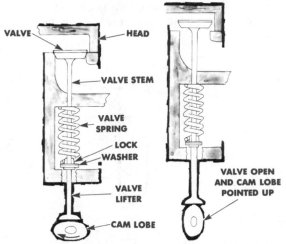

Figure 7-9: Valve lifters are metal rods that push on the valve stems. The cam lobe opens the valve and the spring pulls it closed.

Valve springs are used to close the valves. In normal operation, the spring holds the valve closed. When the cam lobe pushes the valve open it compresses the spring. As the lifter drops, the valve is pulled closed by the spring.

The valve stem passes through the valve spring coils. A washer is attached to one end of the stem. The end of the spring rests on this washer. A keeper, or lock, holds the washer on the valve stem (figure 7-11). The other end of the coil spring rests on the engine block.

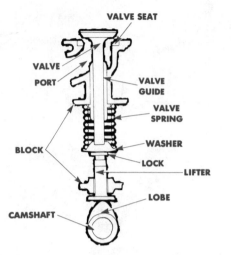

Figure 7-10: This diagram shows all the parts used to open and close the valves of a small engine.

There is one other important part of the valve train. It is called the valve guide. The valve guide is a small tube that fits into the cylinder. It keeps the valve centered over the valve seat.

Figure 7-10 shows the parts that are common to the valve train of all four-stroke cycle engines:

● Intake valve
● Exhaust valve
● Camshaft
● Valve Lifters
● Valve springs
● Valve guides

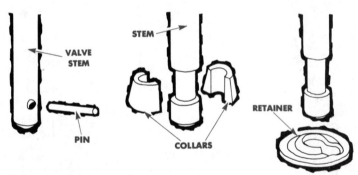

Figure 7-11: Pins, collars, and retainer rings are used to hold the washers to the ends of the valve stems. The washers hold the valve springs in place.

SELF CHECK

1. What are the two valves in each cylinder called?
2. How does the camshaft work?
3. What closes the valves?
4. How does a cam lobe work?

Activities for Unit 7

1. Using a cut-away engine model, identify the parts of the valve train.
2. Make a sketch of the valve train of a small engine and name all the parts.
3. Turn the crankshaft on a cut-away engine model and describe the operation of the valves.

Engines are measured by how big they are on the inside. The three main measurements that describe an engine's size are:

● Bore
● Stroke
● Displacement

The <u>bore</u> is the diameter of the cylinder. Diameter is the distance from one side of the cylinder to the other across the middle (figure 8-1). A 2 to 3 inch bore is common on small engines. Car engines often have a bore between 3 and 4 inches.

<u>Stroke</u> is the second measurement. The stroke is the distance the piston moves from the top of the cylinder to the bottom. This stroke may be 2 to 3 inches in small engines. A car engine stroke may be 4 inches.

<u>Displacement</u> is the volume the piston fills or empties in the cylinder (figure 8-2). It is measured in cubic inches or cubic centimeters. Cubic centimeters are called cc's. In this book we will use cubic inches.

A bigger bore or longer stroke increases displacement. A smaller bore or shorter stroke reduces displacement. If the engine has more than one cylinder, the displacement of all its cylinders are added together to get the total displacement.

Small engines may have 3 to 15 cubic

Figure 8-1: The bore is the diameter (distance across the center) of the cylinder.

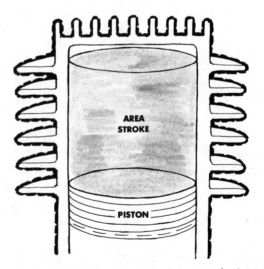

Figure 8-2: Displacement is the area (volume) of the cylinder when the piston is at the bottom of its stroke.

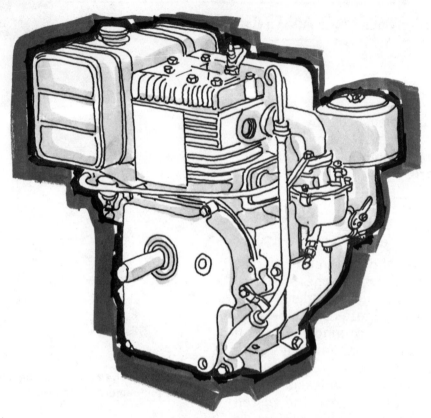

Figure 8-3: In order to measure displacement, all the engine's main parts must be present.

inches of displacement. Automotive engines may have more than 400 cubic inches of displacement.

To measure displacement you need a small engine with all its parts: piston, crankshaft, connecting rod, cylinder head, etc. (figure 8-3). You also will need a combination wrench (figure 8-4) that will fit the head bolts of the engine.

Have a can of oil with a pouring spout (figure 8-5), a funnel (figure 8-6), and a measuring tube (figure 8-7) at hand.

Start by removing the spark plug and loosening the cylinder head bolts with the combination wrench (figure 8-8). Turn them to the left to loosen. Remember that "left is loose", and "right is tight". Take the head bolts out of their holes, and lift off the cylinder head (figure 8-9).

Turn the crankshaft with your hand. You can see the piston go up and down in the open cylinder. Keep turning the crankshaft until the piston is at the bottom of the cylinder. Make sure the engine is level.

Figure 8-4: A combination wrench is used to remove the cylinder head.

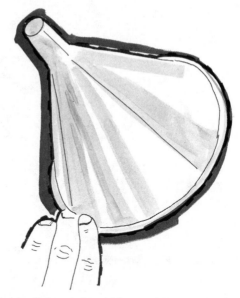

Figure 8-6: A funnel is used for pouring the oil.

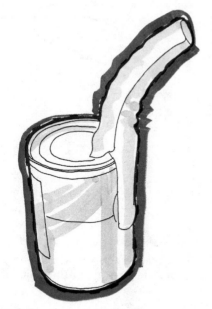

Figure 8-5: A can of oil with a pour spout.

Figure 8-7: A measuring tube.

Fill the cylinder to the top with oil (figure 8-10).

Turn the cylinder head upside down.

It must be level and the spark plug must be in it. Fill the combustion chamber with oil (figure 8-11).

Put the funnel into the measuring

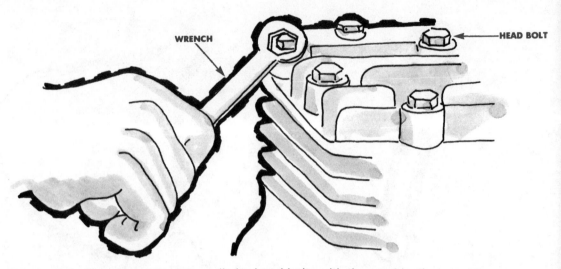

Figure 8-8 Remove each of the cylinder head bolts with the combination wrench.

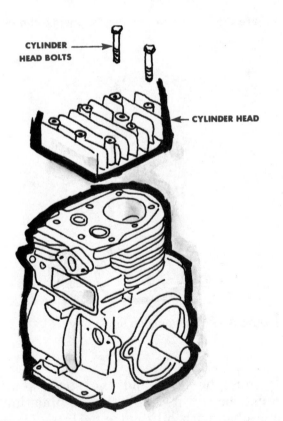

Figure 8-9: Remove the cylinder head.

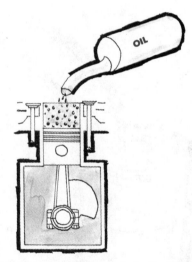

Figure 8-10: Turn the crankshaft until the piston is at the bottom of its stroke. Fill the cylinder with oil.

tube, and pour the oil from the cylinder head into the funnel. Pour the oil in the cylinder into the funnel also. The measuring tube will show how much oil

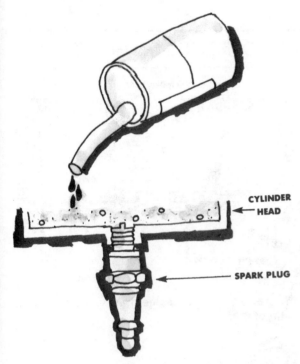

Figure 8-11: Turn the cylinder head over and fill it with oil.

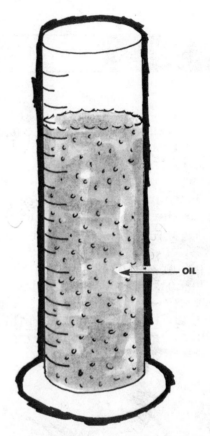

Figure 8-12: Pour the oil from the cylinder head and cylinder into the measuring tube. The amount of oil is equal to the cubic displacement.

was in the cylinder and cylinder head (figure 8-12).

If there are eight cubic inches of oil, that is the cylinder's displacement. Add the displacement of all cylinders together for total engine displacement.

SELF CHECK

1. Define the term "bore".
2. Define the term "displacement".
3. How does a shorter stroke change displacement?
4. Define the term "stroke".

Activities for Unit 8

1. Using a ruler or scale, measure the bore of a small engine.

2. Using a ruler or scale, measure the stroke of a small engine.

3. Measure the displacement of several shop engines to find out which one has the largest displacement.

Compression ratio is a measure of how much the piston squeezes the air-fuel mixture on the compression stroke.

In the compression stroke, the piston rises to the top of the cylinder. The air-fuel mixture is squeezed into the combustion chamber (figure 9-1). This is important, because higher compression means the engine develops more power.

To find the compression ratio, first measure the cylinder displacement as

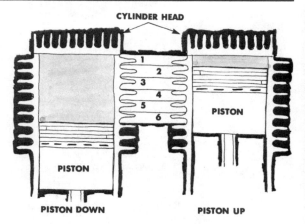

Figure 9-2: This engine has a compression ratio of 6 to 1.

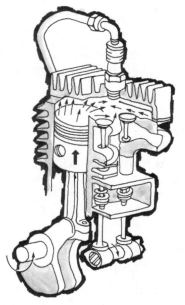

Figure 9-1: The piston rises in the cylinder on the compression stroke and compresses the air-fuel mixture into the combustion chamber.

shown in Unit 8. Then, with the piston at the top of its stroke, measure how much displacement is left.

In figure 9-2, with the piston at the bottom of its stroke, there are 6 cubic inches of displacement in the cylinder. At the top of its stoke there is only 1 cubic inch of space left. So the compression stroke squeezes 6 cubic inches of air-fuel mixture into 1 cubic inch of space. This is a compression ratio of 6 to 1.

If the displacement was 10 cubic inches with the piston at the bottom of its stroke, and 1 cubic inch was left when the piston reached the top, the compression ratio would be 10 to 1.

Most small engines for lawnmowers, edgers, and minibikes have a compression ratio of about 6 to 1. Engines in

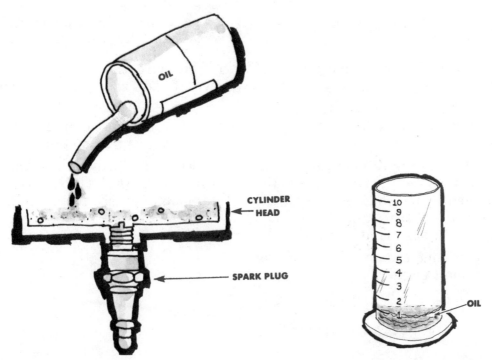

Figure 9-3: Measure the displacement in the cylinder head. In this illustration the oil from the cylinder head measures 1 cubic inch.

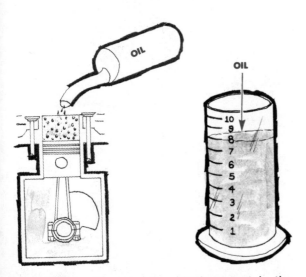

Figure 9-4: Measure the displacement in the cylinder. In this illustration it is 8 cubic inches.

cars and motorcycles have compression ratios of about 8 to 1.

To measure compression ratio you need the same things you used to measure displacement: engine, oil, funnel, and measuring tube.

Begin by measuring the displacement in the cylinder head the same way you did in Unit 8 (figure 9-3). Write down its displacement. Next, measure the displacement of the cylinder when the piston is at the bottom of its stroke (figure 9-4), and write down its displacement.

If the total cylinder displacement is 8 cubic inches and the cylinder head displacement is 1 cubic inch, the compression ratio is 8 to 1.

SELF CHECK

1. What does the term "compression" mean?

2. Does a high compression or a low compression engine have more power?

3. What is the compression ratio of an engine with 7 cubic inches of displacement when the piston is at the bottom of its stroke and 1 cubic inch when the piston is at the top?

4. What is the compression ratio of most small engines?

Activities for Unit 9

1. Measure the compression ratio of several shop engines to find out which has the highest compression.

2. Look up the compression ratio of your family's automobile.

3. Find out which has the highest compression ratio your family automobile or lawnmower.

IGNITION

The method used to start the air-fuel mixture burning in the engine cylinder is called <u>ignition.</u> The air-fuel mixture ignites and starts burning. The pressure caused by the burning pushes the piston down.

Most small engines use a magnet to ignite the fuel-air mixture.

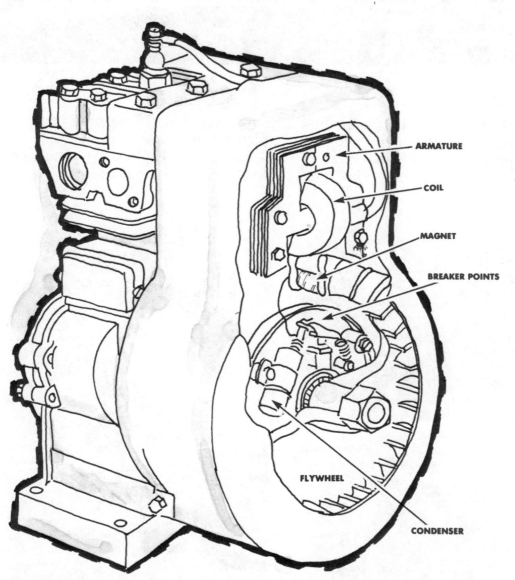

ARMATURE

COIL

MAGNET

BREAKER POINTS

FLYWHEEL

CONDENSER

Figure 10-1: The basic parts of a magneto are: the magnet, the armature, the coil, the breaker points, and the condenser.

An engine develops power by burning the air-fuel mixture. The burn will not start by itself. It is done by an electric spark which is made by a magneto. This spark is called the ignition spark.

The basic parts of a magneto (figure 10-1) are:

- Magnet
- Armature
- Coil
- Breaker points
- Condenser

The word magneto comes from the word magnet. All magnetos have magnets. Some metals stick to a magnet. This is called magnetism. Magnetism is used to make electricity.

The magneto magnet is attached to the engine's flywheel (figure 10-2). There are many ways to attach it. The main thing is that the magnet turns when the flywheel turns.

The armature is made from several strips of soft iron. The strips are squeezed together tightly.

The coil is made with two wires. One wire, the heaviest, is called the primary wire. One of its ends is attached to the armature. The other end is attached to the breaker points.

The secondary wire is the second wire. It is thinner than the primary wire. It is wrapped around the primary wire. One end of the secondary wire is attached to the armature. The other end attaches to the thick wire that is

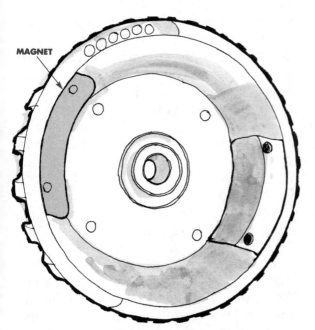

Figure 10-2: The moving part in a magneto is the magnet. It attaches to the flywheel.

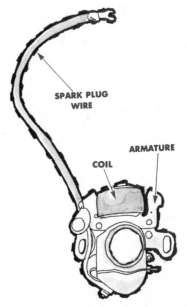

Figure 10-3: The magnet must pass near the armature and coil.

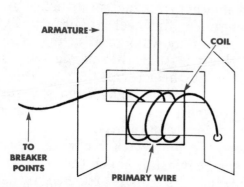

Figure 10-4: The armature is made of several iron strips squeezed together. The primary wire of the coil wraps around the armature. One end attaches to the armature. The other end attaches to the breaker points.

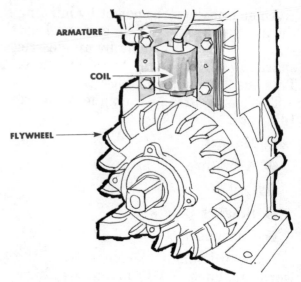

Figure 10-6: This is one way to mount the armature and coil next to the flywheel.

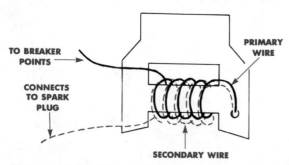

Figure 10-5: The secondary wire in the coil wraps around the primary wire. It also attaches to the armature. The other end goes to the spark plug.

called the spark plug wire. The spark plug wire is attached to the spark plug (figure 10-5).

The coil and armature are mounted so the magnet passes under them as the flywheel turns (figure 10-6). As the magnet moves past the armature the

magnetic force passes from the magnet through the coil into the armature, and back to the magnet again (figure 10-7). As the magnet continues to pass under the armature, the direction of the magnetic force changes (figure 10-8). Now the magnetic force enters the armature from the other end and flows in the opposite direction. This is important because, when the magnetic flow changes direction, a small amount of electricity is produced in the primary wire of the coil.

There is still not enough electricity to make a spark to ignite the air-fuel mixture in the cylinder. This is done in the secondary coil. Unit 11 explains how low voltage electricity in the primary wire is changed to high voltage electricity in the secondary wire.

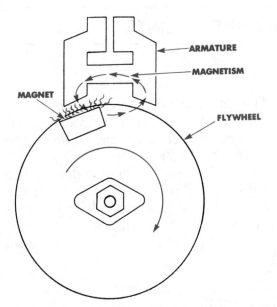

Figure 10-7: Magnetism from the magnet flows through the armature and returns to the magnet.

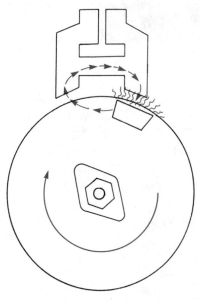

Figure 10-8: When the magnet travels far enough, the magnetism reverses its direction in the armature.

SELF CHECK

1. Why does an engine need an electrical spark?

2. What engine part makes the spark?

3. List six basic parts of a magneto.

4. What are the two wires inside a coil called?

Activities for Unit 10

1. Using a cut-away engine model, identify the magneto, the armature, and the coil.

2. Find the magnets on a small engine flywheel.

3. Turn a small engine flywheel past an armature and explain how electricity is developed.

Breaker points are round metal pieces that touch each other so electricity may flow through them. They are a switch that is opened and closed by the crankshaft (figure 11-1).

Breaker points have six working parts:

- Breaker arm
- Arm support
- Plunger
- Spring
- Moving point
- Stationary point

The moving point is the point attached to the breaker arm. A small round rod called a plunger moves the arm (figure 11-2).

One end of the plunger rides on the crankshaft. There is a flat spot on the crankshaft. When the plunger rides over the flat spot it lowers the breaker arm. The spring pulls the points closed (figure 11-3).

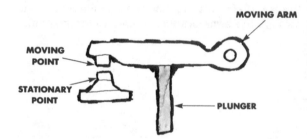

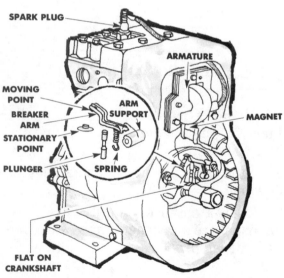

Figure 11-1: By locating the breaker points near the crankshaft, in this case behind the flywheel, it is possible to open and close the points with the crankshaft.

Figure 11-2: The parts of the breaker points are: the stationary point, the moving point, the moving arm, and the plunger.

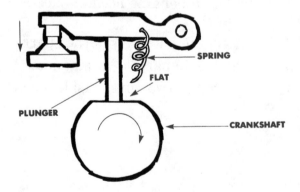

Figure 11-3: A flat spot on the crankshaft lowers the plunger and closes the points.

Figure 11-4: As the crankshaft turns, the rounded part raises the plunger and opens the points.

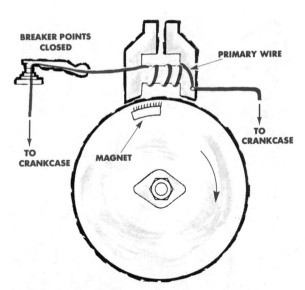

Figure 11-6: When the points are closed, electricity flows through the primary wire and the closed points into the crankcase.

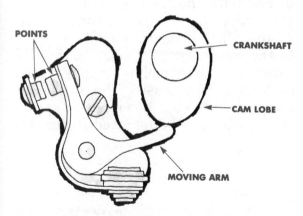

Figure 11-5: This set of points is opened by a cam lobe on the crankshaft. It has no need for a plunger because of the shape of the moving arm.

As the plunger rides past the flat spot on the crankshaft it opens the breaker points (figure 11-4).

Instead of a plunger and flat spot on the crankshaft, some engines have a cam lobe on the crankshaft that opens the points (figure 11-5).

As explained in Unit 10, when the magnet on the flywheel passes the armature, a small amount of electricity is made in the primary wire. This electricity flows to the moving point. If the points are closed, the electricity can flow through them into the crankcase (figure 11-6).

The turning crankshaft raises the plunger which opens the points. Open points break the circuit. The electricity flowing in the primary wire stops. This causes a magnetic flow in the secondary wire. Such a magnetic flow creates high voltage electricity. This voltage may be as high as 20,000 volts. It is enough electricity to create a spark that can ignite the air-fuel mixture in the combustion chamber (figure 11-7).

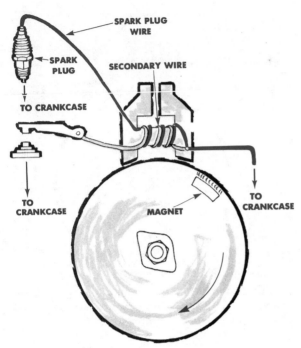

Figure 11-7: Open points stop the flow of electricity in the primary circuit. This creates a magnetic surge which causes a high voltage electricity to flow in the secondary wire.

On most small engines the breaker points are located behind a cover called the breaker point cover (figure 11-8). This cover protects the points from dirt and water.

Two general styles of breaker points are used on most small engines:

● Plunger operated

● Cam operated

Plunger operated breaker points are opened and closed by a flat spot on the rotating crankshaft (figure 11-9). The condenser and stationary point are combined into one part.

Replace this type of breaker point set by removing the movable point, and the stationary point and condenser assembly. The primary wire is attached to the condenser by a small spring. A small depresser tool is supplied with each new condenser. The depressor tool is used to push down on the spring so the wire can be removed (figure 11-10).

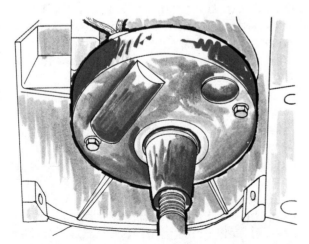

Figure 11-8: The breaker point cover must be removed to service the points.

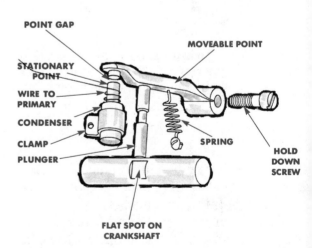

Figure 11-9: The contact points are opened by a plunger.

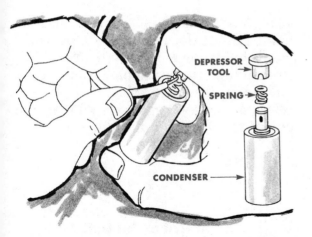

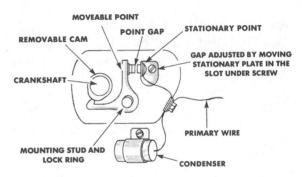

Figure 11-10: Removing the primary wire from the condenser.

Figure 11-12: Some breaker points are opened by a cam.

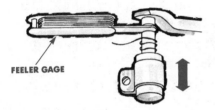

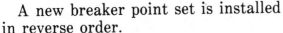

Figure 11-11: Adjusting the breaker points by moving the condenser.

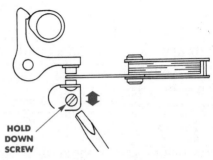

Figure 11-13: The breaker points are adjusted by moving the stationary point.

A new breaker point set is installed in reverse order.

The width of the gap between the points must be set accurately. It determines coil build-up time and sets the engine's timing. Turn the crankshaft until the points are open to their widest gap. Select a feeler gage of the recommended size (usually .030″) and position it between the two points as shown in figure 11-11. Adjust the gap by loosening the lock screw holding the condenser and moving the condenser up or down. The gap is set correctly when the feeler gage will slide in and out of the gap with a light drag.

Cam operated breaker points (figure 11-12) are opened and closed by a removable cam attached to the crankshaft. The condenser is separate from the breaker points. Remove the spring clip that holds the point cover in place. Lift off the point cover. Disconnect the primary and condenser wires. Remove the movable point by removing the lock ring and lifting the point assembly off the mounting stud. Remove the stationary point and condenser, and you're ready to install the new breaker point set. The new breaker points are installed in reverse order.

Rotate the engine until the high part of the cam contacts the movable point arm. A feeler gage is used to adjust the point gap by loosening the stationary point hold down screw (figure 11-13) and sliding the stationary point back and forth. The gap is adjusted correctly when the feeler gage will slide in and out of the gap with a light drag.

SELF CHECK

1. How are the points on a small engine opened and closed?
2. What do the points do?
3. What are the working parts of the breaker points?
4. What does the plunger do?

Activities for Unit 11

1. Find the breaker points on a small engine and explain how they are opened and closed.
2. Remove and replace a set of breaker points.
3. Adjust the gap of a set of breaker points.

When the points are open, the electricity in the primary wire must still try to go somewhere. The condenser is used to store the electricity from the primary wire while the points are open and it has nowhere to go. Without the momentary storage allowed by the condenser the electricity in the primary wire would jump across the open points (figure 12-1). This would create a spark that would allow the electricity in the primary wire to continue flowing. The circuit would not be broken. If the circuit was not broken the magnetic field around the primary wire could not collapse, and no high voltage would be formed in the secondary wire to create the spark that ignites the air-fuel mixture.

The condenser is a small electrical part that looks like a tiny can. It is made of two long sheets of foil separated by sheets of paper. The foil and paper are rolled into a tight roll. Electricity can flow into the foil. The large area of the foil allows room to store a lot of electricity (figure 12-2). The condenser is attached to the magneto by a bracket. A wire from one end of the condenser is hooked to the moving point (figure 12-3). Electricity travels

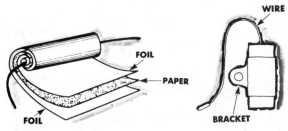

Figure 12-2: There is only one wire to the condenser. It is attached to sheets of foil inside the condenser.

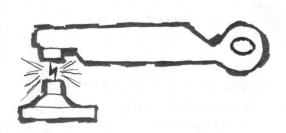

Figure 12-1: When the breaker points open, electricity tries to jump across the points, reducing the high voltage electricity and burning the points.

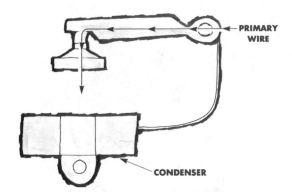

Figure 12-3: Electricity does not flow to the condenser when the points are closed. It flows from the primary wire, through the points, and into the block.

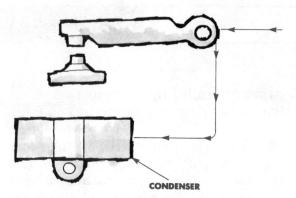

Figure 12-4: When the points are open, the electricity flows to the condenser where it is stored.

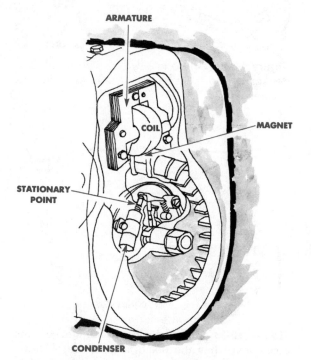

Figure 12-5: Sometimes the condenser and stationary point are one piece.

in this wire to the condenser (figure 12-4).

Without a condenser, the electricity would jump across the open points. A side effect of electricity jumping the points is that the points would be burned until they wouldn't work at all.

Some engines have the condenser and stationary point in one piece. When this occurs, the primary wire goes to the stationary point and the moving point is attached to the block (figure 12-5).

SELF CHECK

1. What happens if electricity jumps across the breaker points?
2. What causes burned breaker points?
3. What does the condenser do with electricity?
4. How does a one piece condenser-stationary point work?

Activities for Unit 12

1. Locate the condenser on a small engine and explain how it works.
2. Remove and replace a one piece condenser-stationary breaker point assembly.
3. Remove and replace a condenser that is connected to the points with a wire.

The special wire that carries the high voltage electricity spark plug wire made by the magneto into the combustion chamber is connected to a spark plug (figure 13-1). The spark plug ignites the air-fuel mixture in the combustion chamber.

The main parts of a spark plug are (figure 13-2):

● Terminal

● Insulator

● Shell

● Center electrode

● Ground electrode

The terminal is connected to the magneto by the spark plug wire. High voltage electricity flows from the magneto to the terminal of the spark plug wire.

The center electrode is a solid wire. It goes from the terminal to the bottom of the spark plug. It is surrounded by the insulator and the shell. The insulator is made of ceramics. Ceramics do not allow electricity to pass through them. The shell is made of metal and has threads so it can be screwed into the cylinder head.

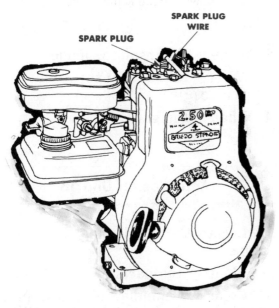

Figure 13-1: The spark plug in the cylinder head is connected to the magneto by a spark plug wire.

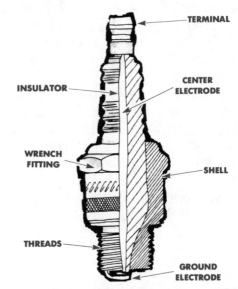

Figure 13-2: The main parts of the spark plug are the terminal, the insulator, the shell, the center electrode and the ground electrode. Notice that the top of the shell is made to fit a wrench and the bottom is threaded to screw into the cylinder head.

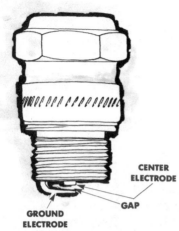

Figure 13-3: The space between the center electrode and the ground electrode is called the gap.

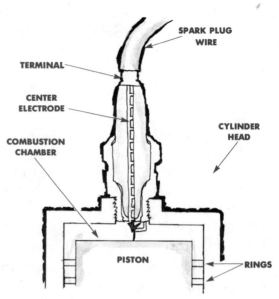

Figure 13-4: High voltage electricity travels from the magneto to the terminal of the spark plug along the spark plug wire. It continues down the center electrode to the gap which it jumps, causing a spark. This ignites the air-fuel mixture and starts the power stroke in the cylinder.

The last part of the spark plug is the ground electrode. It is a flat wire that attaches to the metal shell and forms a gap with the center electrode (figure 13-3).

Notice that the center electrode and ground electrode look like a pair of open points. The distance between them is called the gap. High voltage electricity jumps this gap, causing a spark that ignites the air-fuel mixture in the combustion chamber to ignite (figure 13-4).

2. What is the insulator made of?

3. Where is the center electrode located?

4. What is the gap?

Activities for Unit 13

1. Locate the spark plug of a small engine.

2. Identify the parts of a spark plug.

3. Collect a number of spark plugs from around your school shop. What differences are there between the spark plugs?

SELF CHECK

1. How does electricity get from the magneto to the spark plug?

When an engine will not start, the first thing to check is the spark. If something is wrong with the magneto there is no spark.

A spark tester (figure 14-1) is used to test the magneto for spark. When checking any engine for spark, do not touch the end of the spark plug wire when the flywheel is turning. The high voltage electricity carried by this wire is dangerous.

First take the spark plug wire off the spark plug by pulling it off the terminal. Attach the free end of the spark plug wire to the spark tester. Touch the other end of the spark tester to the cylinder head (figure 14-2).

Spin the flywheel. This should cause a spark to jump the tester gap (figure 14-3). If there is no spark, the fault is in the magneto.

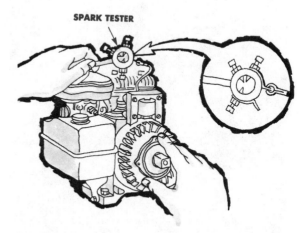

Figure 14-2: The mechanic tests for spark by attaching the spark plug wire to the spark tester and grounding the spark tester on the cylinder head.

Figure 14-3: High voltage electricity from the spark wire will jump the gap, making a visible spark if the magneto is working. However, the tester must be grounded on the block for the spark to occur.

Figure 14-1: If the magneto is working, the spark tester will create a spark similar to the one made by a spark plug.

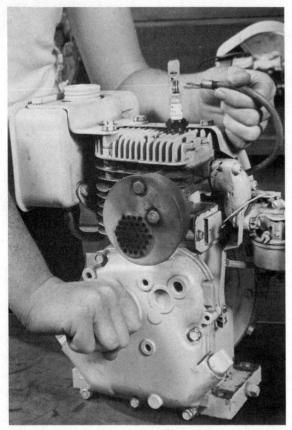

Figure 14-4: To test for spark without a spark tester, hold the spark plug wire terminal end near the spark plug terminal. Spin the flywheel. A spark should jump from the wire to the spark plug terminal.

There is another way to check for spark. Grasp the spark plug wire several inches from the end. Hold the terminal end of the wire close to the spark plug terminal (figure 14-4). Spin the flywheel again. A spark should jump from the wire to the spark plug terminal. If it doesn't, the magneto is not working.

Five common problems with the magneto are:

- A broken key that holds the flywheel on the crankshaft
- Too much gap between the breaker points
- Too little gap between the breaker points
- Dirty or burned breaker points
- A primary wire that is not connected right

SELF CHECK

1. What is the first thing you should check when an engine won't start?
2. Why shouldn't you touch the spark plug wire terminal when the crankshaft is turning?
3. How would you test a spark if you did not have a spark tester?
4. List five reasons a magneto might not be working.

Activities for Unit 14

1. Make a list of the steps you would follow to find out if an engine has spark.

2. Test an engine for spark with a spark tester.

3. Test an engine for spark without a spark tester.

If the magneto works and the engine won't start, the trouble may be the spark plug.

Spark plugs normally have three problems:

● The center and ground electrodes are burned by the spark

● Burned oil and gasoline may have filled the gap between electrodes

● The insulator may have cracked or the center electrode may be damaged

Use a spark plug socket to remove the spark plug from the engine. The spark plug socket (figure 15-1) fits over the spark plug. A "T" handle on the end of the socket is used to unscrew the spark plug (figure 15-2).

Look at the electrode end of the spark plug. If it is dirty it must be cleaned (figure 15-3). Use a wire brush on the electrode area (figure 15-4). Brush until the electrodes are clean (figure 15-5).

Check the center electrode. If the

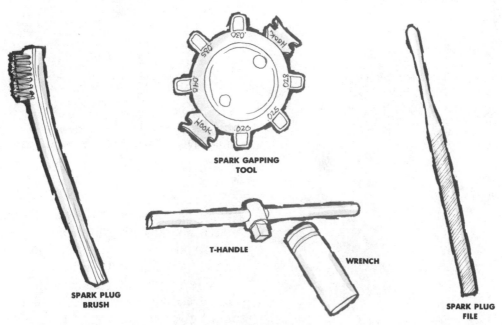

SPARK GAPPING TOOL

SPARK PLUG BRUSH

T-HANDLE

WRENCH

SPARK PLUG FILE

Figure 15-1: These tools are used to clean and gap spark plugs. Spark plug sockets on "T" handle wrenches are used to remove and replace spark plugs.

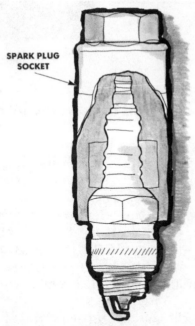

SPARK PLUG
SOCKET

Figure 15-2: The spark plug socket must fit all the way down to the special fitting on the spark plug shell. Most spark plugs are 13/16 of an inch.

Figure 15-4: Wire brush the electrodes until the metal shines with no coating.

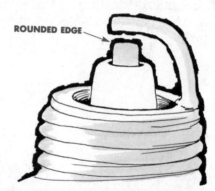

ROUNDED EDGE

Figure 15-5: This spark plug has been wire-brushed clean, but the edge of the center electrode has been eaten away by the spark, rounding it.

Figure 15-3: Dirty spark plug electrodes are usually covered with a grey or black ash. Sometimes the gap is completely filled.

edge is rounded, file the electrode flat with a spark plug file (figure 15-6). The electrode should look like the one shown in figure 15-7.

The spark plug gage gapping tool is used to set the gap distance between the center and ground electrodes (figure 15-8). Each wire on this tool is a different size. The thickness of the

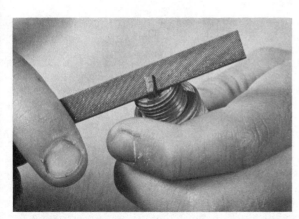

Figure 15-6: This is how a center electrode is squared with a spark plug file.

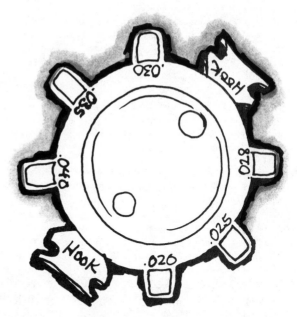

Figure 15-8: The gapping tool has different sized wires around its edges.

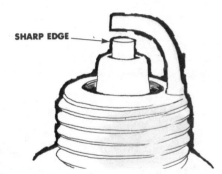

Figure 15-7: A cleaned and filed spark plug has square edges on the electrodes and bright metal.

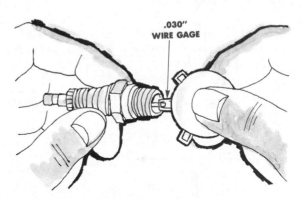

Figure 15-9: The spark plug gap is measured and adjusted with a gapping tool.

wire is marked on the tool. Each wire measures a different spark plug gap. The hook on the spark plug gage is used to bend the ground electrode. This adjusts the spark plug gap.

A repair manual for the engine will tell you how wide to set the gap. Find the gap tool wire that is the called for size. Most small engines use a gap of .030 (thirty thousandths) of an inch.

Slide the wire into the space between the electrodes. It should fit snugly

(figure 15-9). If the wire will not go in, the gap is too small. If the wire fits loosely, the gap is too big.

Use the hook on the spark plug gapping tool to make the gap smaller or

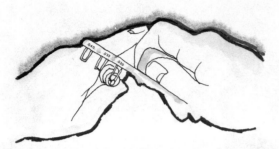

Figure 15-10: Gapping tools come in different shapes. After regapping, always check the new gap.

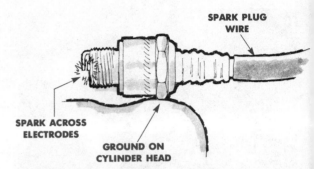

Figure 15-11: A spark plug may be defective inside. It can be tested by grounding the spark plug shell on the cylinder head, connecting the spark plug wire and spinning the flywheel. A good spark plug will show a spark across the gap.

larger. Bend the ground electrode closer to the center electrode to make the gap smaller. Bend the electrode away from the center electrode to make the gap bigger. Always recheck the new gap (figure 15-10).

If the electrodes are okay, it is possible the spark plug is no good. The spark plug may be checked by attaching the spark plug wire to the terminal. Ground the spark plug on the cylinder head. Then spin the flywheel (figure 15-11).

If there is no spark between the electrodes, replace the spark plug with a new one.

Screw the clean or new spark plug into the cylinder head by hand. Turn it to the right. Do not use a wrench until you can no longer turn the spark plug by hand. One-half turn with the socket is enough. More turns with the wrench may damage the threads.

SELF CHECK

1. List three most common spark plug problems.
2. What should be done to a rounded electrode?
3. How is a spark plug cleaned?
4. How is the gap measured?

Activities for Unit 15

1. Remove a spark plug from an engine.
2. Clean a spark plug.
3. Regap a spark plug.

CARBURETION

3

An engine must have a mixture of fuel and air to run. Getting the right amount of fuel and air mixed together and into the engine is called carbure-tion. The engines we study in power mechanics all have some kind of carburetion system.

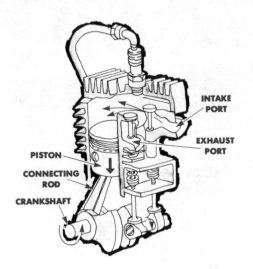

PISTON

CONNECTING ROD

CRANKSHAFT

INTAKE PORT

EXHAUST PORT

Figure 16-1: On the intake stroke the piston moves down. Due to reduced air pressure it draws air into the cylinder through the intake valve port.

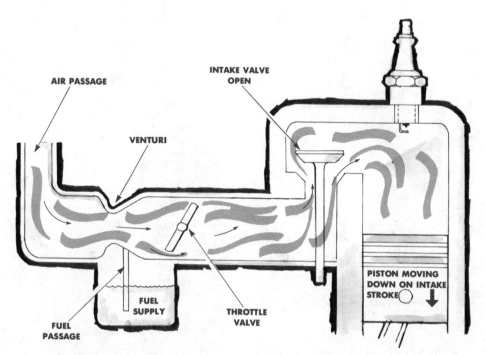

AIR PASSAGE

VENTURI

INTAKE VALVE OPEN

FUEL SUPPLY

FUEL PASSAGE

THROTTLE VALVE

PISTON MOVING DOWN ON INTAKE STROKE

Figure 16-2: Air is pulled through the carburetor before it enters the engine. The carburetor is basically a tube attached to the engine at one end and open at the other end.

The carburetor mixes and meters fuel and air before they enter the engine. There are many different types of carburetors. But their basic operation is the same.

On the intake stroke, the piston draws air into the cylinder (figure 16-1). This air is pulled through the carburetor. It enters the cylinder at the intake valve.

The carburetor (figure 16-2) is shaped like a tube attached to the engine. A very basic carburetor would have five parts:

● An air passage into the engine
● A fuel supply
● A passage for fuel to get into the air passage
● A venturi
● A throttle valve

Air is pulled into the open end of the air passage tube on the intake stroke.

The venturi is a narrow place in the tube. As the air is drawn through this narrow place, it moves faster. The result is a lower pressure in the venturi area of the carburetor (figure 16-3).

A small amount of fuel is stored in a bowl under the venturi. A pick-up tube runs from the fuel to the venturi (figure 16-4).

The low pressure at the venturi pulls fuel up the tube. The fuel mixes with

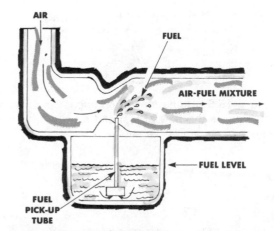

Figure 16-4: The fuel pick-up tube is in the low pressure area (venturi). Fuel is drawn up the tube by the low pressure (vacuum). It mixes with the fast moving air.

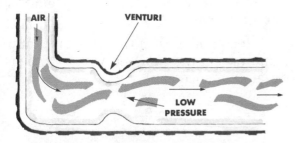

Figure 16-3: The venturi is a narrow place in the carburetor. The air moving through the carburetor is forced to move faster as it passes through the venturi. This creates a low pressure area on the engine side of the venturi.

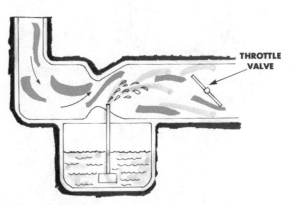

Figure 16-5: The throttle valve completes the basic carburetor.

the air moving through the passage at the venturi. Then both air and fuel enter the engine.

The <u>throttle valve</u> controls the engine speed. More fuel and air make the engine run faster. Less fuel and air make the engine run slower (figure 16-

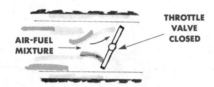

Figure 16-6: In its closed position, the throttle valve slows the flow of fuel and air reaching the engine. This slows the engine.

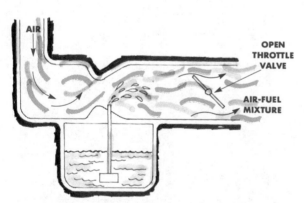

Figure 16-7: Opening the throttle valve lets more air-fuel mixture reach the cylinder so the engine will run faster.

5). The throttle valve is a round plate. It fits inside the carburetor tube. On most small engines it is controlled by a lever mounted near the operator.

When the throttle valve is closed, only a small amount of air and fuel can get into the engine (figure 16-6) and the engine runs slower. Open the valve and the engine will run faster (figure 16-7).

SELF CHECK

1. What does a carburetor do?
2. How does a venturi work?
3. What is the fuel pick-up tube?
4. How does a throttle valve work?

Activities for Unit 16

1. Make a sketch of a basic carburetor and name all the parts.

2. Using a shop carburetor, identify the basic carburetor parts.

3. Open and close a throttle valve on a running engine and see what happens to engine speed.

Chokes make carburetors work better when they are cold. They help to start an engine when it is cold. The choke is a small round valve. It fits in the open end of the carburetor (figure 17-1).

Closing the choke cuts down on the flow of air and increases the vacuum. On an intake stroke with the throttle open and the choke closed the engine gets a rich mixture (lots of fuel and little air). A rich mixture will ignite better when the engine is cold, better than the usual mixture of air and fuel. A level mounted on the carburetor opens and closes the choke (figure 17-2).

The carburetor has two adjustment screws. They are:

● High speed adjustment screw

● Low speed adjustment screw

The <u>high speed adjustment screw</u> controls the amount of fuel that can go up the fuel pick-up tube. The head of the screw is outside the carburetor. The end of the screw is in the pick-up tube (figure 17-3). When it is screwed in, the high speed adjustment screw limits the fuel that can pass through the pick-up tube at high speeds. It must be adjusted to give the engine the right amount of fuel to run properly at high speeds.

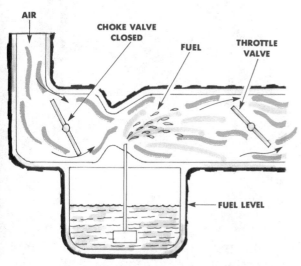

Figure 17-1: Closing the choke valve reduces the amount of air entering the engine.

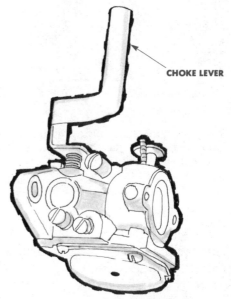

Figure 17-2: A choke lever mounted on the carburetor opens and closes the choke.

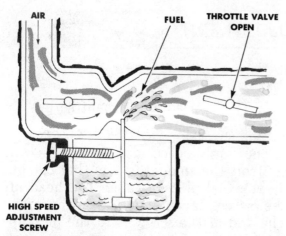

Figure 17-3: The high speed adjustment screw controls the amount of fuel that comes up the pick-up tube.

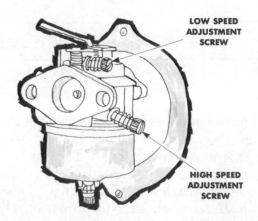

Figure 17-5: In this carburetor, the high and low speed adjustment screws are near each other.

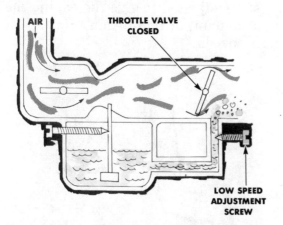

Figure 17-4: The low speed adjustment screw controls the amount of fuel that enters the carburetor behind the throttle valve.

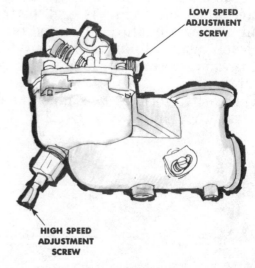

Figure 17-6: Placement of low and high speed adjustment screws varies from carburetor to carburetor.

The low speed adjustment screw also enters the carburetor from the outside. It controls the flow of fuel in a special passageway that enters the carburetor behind the throttle. At low speeds the closed throttle closes the air passage-

way and enough fuel cannot be pulled up the pick-up tube to the venturi to keep the engine running. Extra fuel must be added to the air, but it must be

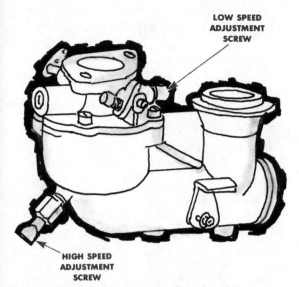

LOW SPEED
ADJUSTMENT
SCREW

HIGH SPEED
ADJUSTMENT
SCREW

Figure 17-7: This carburetor has the high speed adjustment screw on the bottom and the low speed adjustment screw on top.

added behind the throttle. This is done with a special fuel passageway (figure 17-4). When it is screwed in, the low speed adjustment screw controls the amount of fuel added to the air behind the throttle.

Figures 17-5, 17-6 and 17-7 show a few of the many different arrangements for adjustment screws.

SELF CHECK

1. How does the choke work?
2. Why is the choke used?
3. What does the high speed adjustment screw do?
4. What happens when the low speed adjustment screw is turned in?

Activities for Unit 17

1. Find the choke valve on a shop carburetor and explain how it works.
2. Find the low speed adjustment screw on a shop carburetor.
3. Find the high speed adjustment screw on a shop carburetor.

The carburetors described in Units 17 and 18 are essentially vacuum carburetors. They work when fuel is drawn into the carburetor by low pressure or a vacuum. The fuel tank is usually mounted directly under a vacuum carburetor (figure 18-1).

Vacuum carburetors have the following main parts:

- An air opening or intake
- A choke near the air opening
- A fuel pipe
- A fuel filter on the fuel pipe
- A throttle valve

The carburetor is mounted to the engine by two bolts (figure 18-2).

Air entering the carburetor is controlled by the choke. The choke lets in more or less air by opening or closing the passageway.

The throttle valve controls the flow of the air-fuel mixture to the engine. It opens and closes the passageway like the choke.

The fuel pick-up pipe is located right below the throttle valve. It runs from the carburetor to the fuel tank. A fuel filter or small screen on the bottom of the fuel pipe keeps dirt out of the pipe. A small ball in the bottom of the fuel

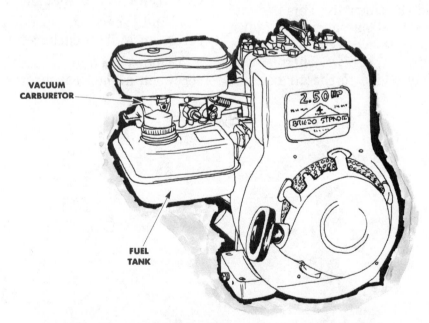

VACUUM CARBURETOR

FUEL TANK

Figure 18-1: Many small engines have vacuum carburetors.

70

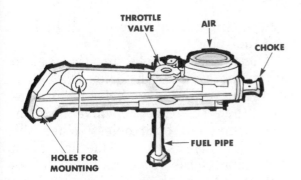

Figure 18-2: In this example of a vacuum carburetor the choke is a sliding cap on the open end of the carburetor.

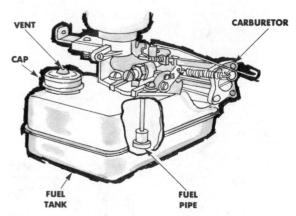

Figure 18-4: Vacuum carburetors are usually mounted on top of the fuel tank.

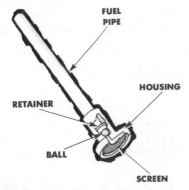

Figure 18-3: When fuel tries to run down the fuel pipe into the fuel tank, it is blocked by a ball which closes the pipe.

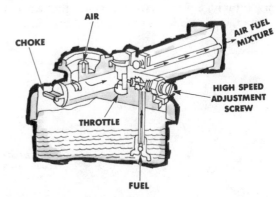

Figure 18-5: The movement of air past the throttle valve causes a vacuum that pulls fuel up the fuel pipe.

pipe allows fuel to run up the pipe, and will not allow it to run down the pipe (figure 18-3).

The <u>fuel tank</u> under the carburetor has a <u>fuel cap.</u> This cap has a vent in it. The vent is a hole which lets air enter the fuel tank as fuel is drawn into the carburetor (figure 18-4).

On the engine intake stroke, the piston moves down in the cylinder. This movement forms a vacuum in the carburetor. Fuel is drawn up the fuel pipe by the vacuum.

The amount of fuel pulled up the fuel pipe is controlled by the high speed adjustment screw (figure 18-5).

The throttle valve squeezes the moving air. This makes the air move faster around the throttle valve. Like the venturi, the low pressure area caused

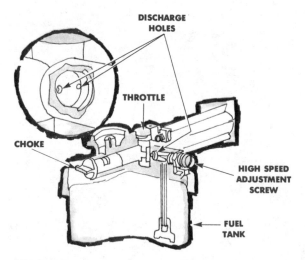

Figure 18-6: Fuel enters through two discharge or metering holes. When the throttle valve is closed, one hole is on each side of the valve.

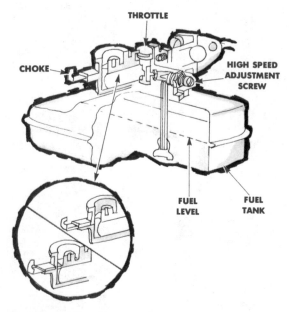

Figure 18-7: The closed choke reduces the amount of air entering the carburetor. With less air, more fuel must be pulled into the engine to fill the vacuum created by the piston.

by the throttle valve helps pull fuel up the fuel pipe.

There are two holes for the fuel to enter the carburetor. They are called metering or discharge holes (figure 18-6). When the throttle valve is open, fuel comes out both holes. When the throttle valve is closed, one hole is on each side of the valve, and fuel comes out of only one hole. This allows the engine to run with the throttle valve closed.

The main reason for the choke is to close off air when the engine is cold. The vacuum created in the carburetor pulls more fuel up the fuel pipe and into the engine (figure 18-7).

SELF CHECK

1. What does the vacuum in a vacuum carburetor do?

2. Why is there a screen in the fuel pipe?

3. Why is there a vent in the fuel tank cap?

4. How does the choke work?

Activities for Unit 18

1. Locate a shop engine with a vacuum carburetor and explain how it works.

2. Remove and replace a vacuum carburetor from an engine.

3. Take a vacuum carburetor apart and identify the parts.

Many small engines use a float carburetor. The float carburetor has a fuel tank that is not attached to it (figure 19-1).

Some fuel tanks are mounted higher than the carburetor (figure 19-2). Gravity makes the fuel flow down into the carburetor.

The float in a float carburetor controls the flow of fuel from the fuel tank to the float bowl. The float bowl is located under the carburetor. The fuel enters the float bowl through an inlet (figure 19-3).

The float lets only a certain amount of fuel into the float bowl (figure 19-4).

Those parts of the float carburetor that make it different from the vacuum carburetor are:

- Float
- Pivot
- Needle valve
- Float bowl
- Inlet

The <u>float</u> is a hollow piece of copper or plastic. It is attached to a <u>pivot</u> by

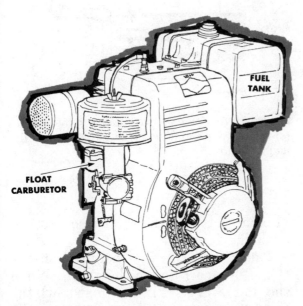

Figure 19-1: An engine with a float carburetor. Note that the fuel tank is mounted on the other side of the engine.

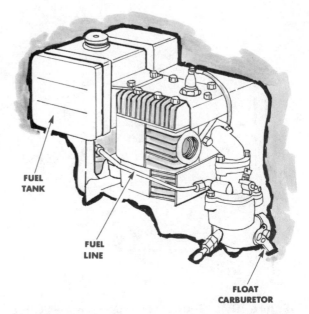

Figure 19-2: A float carburetor is used when the fuel tank is mounted higher than the carburetor.

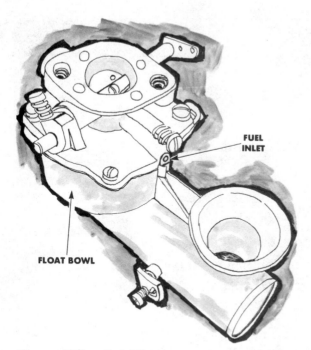

Figure 19-3: Outside view of a float carburetor.

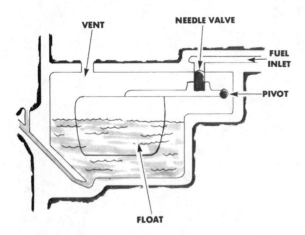

Figure 19-4: The float bowl contains the float, a needle valve, a pivot, a fuel inlet, and an air vent. As the inlet lets fuel into the float bowl, the vent lets air out.

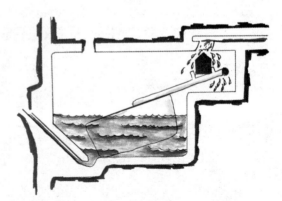

Figure 19-5: The float sinks with the level of the fuel. This opens the inlet and fuel enters the float bowl.

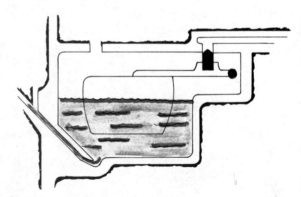

Figure 19-6: When enough fuel is in the bowl to raise the float and needle valve to the inlet, the fuel is shut off.

an arm that lets the float rise and fall with the fuel level.

The needle valve is mounted on the float (or arm) under the inlet. When the fuel level rises in the float bowl, the float rises up and pushes the needle valve into the fuel inlet. As the fuel level in the float bowl drops, the float drops and the needle valve opens the inlet (figure 19-5). When the inlet is open, fuel flows into the bowl.

Fuel is drawn from the float bowl into the carburetor through a tube called a <u>nozzle</u>.

The nozzle works the same as a fuel pipe in a vacuum carburetor. There are different types of float carburetors (figure 19-8), but they all work the same way as vacuum carburetors (figure 19-9). Figures 19-10 through 19-13 show the operation of a float carburetor.

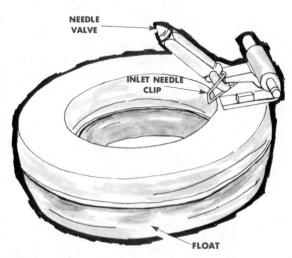

Figure 19-7: This is an example of a float with a needle valve mounted on it.

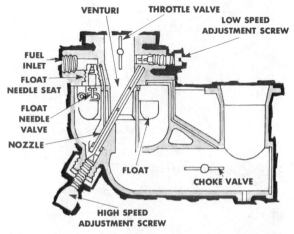

Figure 19-9: The float is the main difference between a float carburetor and a vacuum carburetor.

Figure 19-8: Even though they look different, these two float carburetors work the same way.

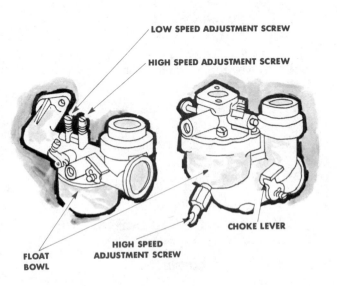

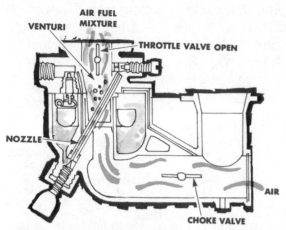

Figure 19-10: Low pressure at the venturi causes fuel to be pulled through the nozzle from the float bowl.

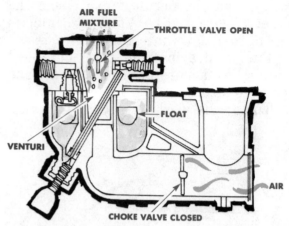

Figure 19-12: A closed choke valve causes more fuel to be drawn from the nozzle into the engine.

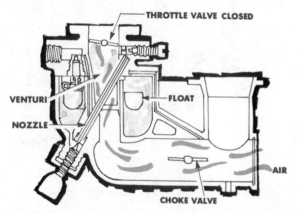

Figure 19-11: When the throttle valve is closed, enough fuel to run the engine comes out of the hole behind the valve.

Figure 19-13: This float carburetor is typical of the kind used on automobile engines.

SELF CHECK

1. What does the fuel line do?

2. Where is the fuel inlet?

3. How do the float and needle valve help control fuel flow?

4. What is the nozzle?

Activities for Unit 19

1. Find a shop engine that has a float carburetor and explain how it works.

2. Remove and replace a float carburetor from an engine.

3. Take a float carburetor apart and name the parts.

Figure 20-1: Some engines, like those on chain saws, have to work upside down. Such enignes use diaphragm carburetors.

Float and vacuum carburetors only work on engines that are used upright. If the engine was turned on its side, the carburetor would not work properly. A chain-saw engine, for example, must work in any position (figure 20-1).

Diaphragm carburetors are made to work in any position. The top part of the diaphragm carburetor is the same as a float or vacuum carburetor (figure 20-2). It has a choke valve, throttle valve, and venturi. It also has a high and low speed adjustment screw (figure 20-3).

Instead of a float bowl, the diaphragm carburetor has a diaphragm.

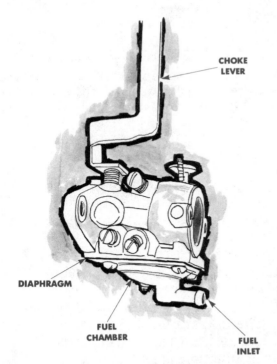

CHOKE LEVER

DIAPHRAGM

FUEL CHAMBER

FUEL INLET

Figure 20-2: The outside view of a diaphragm carburetor is different than the same view of a float carburetor.

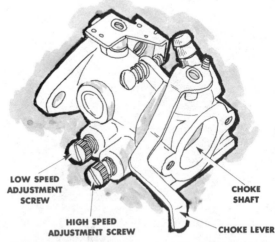

LOW SPEED ADJUSTMENT SCREW

HIGH SPEED ADJUSTMENT SCREW

CHOKE SHAFT

CHOKE LEVER

Figure 20-3: A diaphragm carburetor has a low and a high speed adjustment screw like other carburetors.

The special parts of a diaphragm carburetor are:

- Diaphragm
- Pivot
- Needle valve
- Diaphragm chamber
- Spring
- Control lever

The diaphragm is round. It is made from a flexible rubbery material. This material is stretched across the diaphragm chamber (figure 20-4).

The fuel inlet and the working parts are on one side of the diaphragm. A pocket of air is on the other side.

A small control lever is connected from the center of the diaphragm to a needle valve. The needle valve opens and closes the fuel inlet.

A spring fits between the top of the chamber and the control lever. This spring pushes the needle valve away from the inlet.

Figure 20-5: A chamber full of fuel pushes the diaphragm down. The end of the control lever raises the needle valve to seal the fuel inlet.

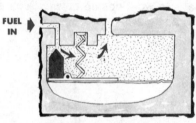

Figure 20-6: When the chamber has little fuel the diaphram is pulled up by the spring. This lets the needle valve open the fuel inlet and fuel flows into the chamber.

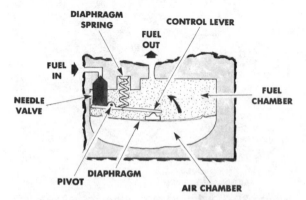

Figure 20-4: Note how the flexible diaphragm expands into the air chamber when the other side of the chamber is full of fuel.

Fuel in the diaphragm chamber pushes the diaphragm down. This raises the needle valve to the inlet and stops the flow of fuel (figure 20-5).

As fuel is used out of the chamber, the spring raises the diaphragm. This lowers the needle valve. Fuel then flows into the chamber (figure 20-6).

The diaphragm will work no matter what position the engine is in. It will even function upside down.

The rest of the carburetor (figure 20-7) is the same as the vacuum or float carburetors.

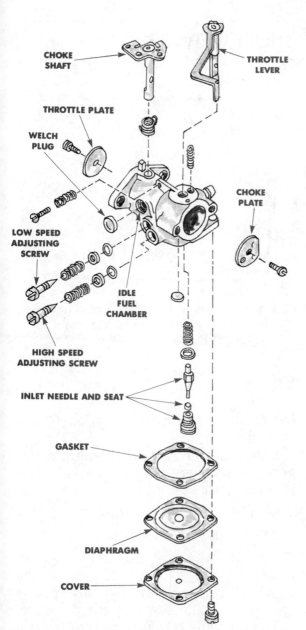

CHOKE SHAFT

THROTTLE LEVER

THROTTLE PLATE

WELCH PLUG

CHOKE PLATE

LOW SPEED ADJUSTING SCREW

IDLE FUEL CHAMBER

HIGH SPEED ADJUSTING SCREW

INLET NEEDLE AND SEAT

GASKET

DIAPHRAGM

COVER

Figure 20-7: An exploded view of a diaphragm carburetor showing all of its parts.

SELF CHECK

1. What kind of engine uses a diaphragm carburetor?
2. Why must a diaphragm be flexible?
3. What does the needle valve do?
4. Where does the control lever connect?

Activities for Unit 20

1. Find a shop engine that has a diaphragm carburetor and explain how it works.

2. Remove and replace a diaphragm carburetor from an engine.

3. Take a diaphragm carburetor apart and name the parts.

The speed of an engine is controlled by the carburetor throttle valve. Open the throttle valve and the engine runs fast. Close the throttle valve and the engine runs slow. Some engines have another speed control. That control is a governor.

A governor holds an engine at a steady speed. An example is a lawn mower engine (figure 21-1). When the lawn mower cuts high grass the engine slows down because it is working harder.

The governor keeps the engine running at a steady speed. It closes the throttle valve if the engine isn't work-

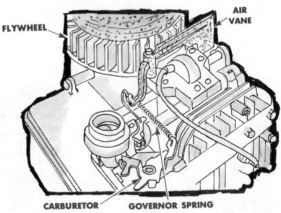

Figure 21-2: This is a typical air vane governor. The faster the flywheel turns, the more air it throws against the air vane.

Figure 21-1: Engines that need to run at a steady speed use governors. In a lawn mower the engine drives a blade which must cut at a steady speed. The governor keeps the blade speed constant.

ing hard. For more power it opens the throttle.

Lawn mower and tiller engines use governors. There are two general types of governors:

● The air vane governor

● The mechanical governor

An air vane governor is a piece of plastic or steel mounted above the flywheel (figure 21-2). This air vane is moved back and forth by air pressure.

The air vane connects to the throttle valve by a piece of wire called a link. A spring is also connected to the link. The spring tries to pull the throttle open (figure 21-3).

The air pressure from the spinning flywheel hits the air vane and tries to

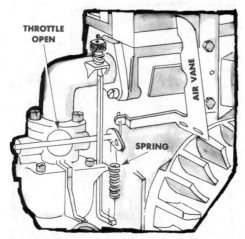

Figure 21-3: With the engine off, the governor spring holds the throttle valve open.

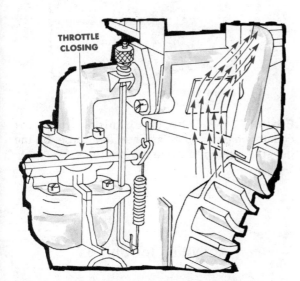

Figure 21-4: Once the engine is running, air pressure against the air vane forces the throttle closed.

close the throttle (figure 21-4) to slow down the engine. The spring works against the vane. If the engine is run-

ning slowly, the air pressure against the air vane is reduced, the spring pulls the throttle open, and the engine speeds up. The spring and vane work together to keep the engine running smoothly.

Part of a <u>mechanical governor</u> is located inside the engine. It is a small gear that rides on the camshaft gear. A set of weights is attached to this small gear (figure 21-5). A governor

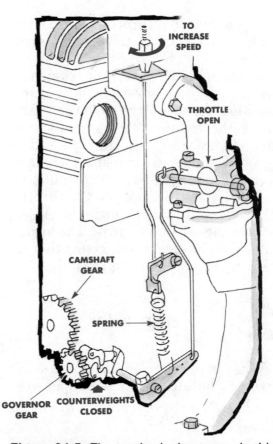

Figure 21-5: The mechanical governor is driven by a camshaft gear that spins the governor gear. Counterweights attached to the governor gear open as the speed increases.

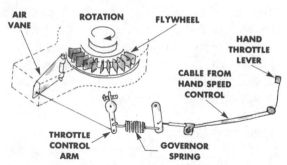

Figure 21-6: The hand throttle sets the governor spring. All control of the engine speed is through the spring.

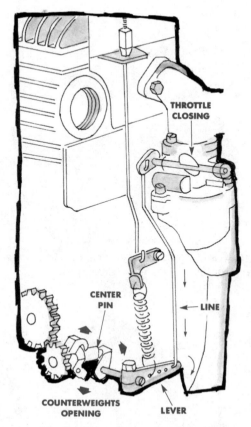

Figure 21-7: The mechanical governor counterweights push against the force of the governor spring to close the throttle.

spring holds the throttle open. The spring is attached to the hand control (figure 21-6).

As the engine runs, the camshaft gear makes the governor gear turn. The high speed of the governor gear causes the counterweights to fly outward (figure 21-7). A pin attached to the weights pushes against an arm and lever. Movement of the lever pulls on the link. Since the link and the throttle are connected, the throttle is closed, and the engine slows down. The spring and counterweights work together to keep the engine running steadily.

SELF CHECK

1. What happens to engine speed when the carburetor throttle is opened?

2. What does a governor do to engine speed?

3. Name the two types of governors.

4. Does the governor spring open or close the throttle?

Activities for Unit 21

1. Find an engine with an air vane governor and name the parts of the governor.

2. Find an engine with a mechanical governor and name the parts of the governor.

3. Look up how to adjust a governor.

On each intake stroke the engine draws in a lot of dirty air. The dirt in this air must not get into the engine cylinder. Dirt in the engine will wear out the cylinder and rings.

Several types of air cleaners are used to remove dirt from the air.

Two of the common air cleaners are:

● Oil-bath air cleaner

● Dry element air cleaner

The <u>oil-bath air cleaner</u> is a can with air passages in it (figure 22-1). Dirty air enters the engine through the top of the can (figure 22-2). The air goes down inside the can. Then it must change direction and start back up again.

There is a pool of oil at the bottom of the can. Dirt in the air is too heavy to make the quick turn. It falls into the oil. Any dirt left in the air is captured

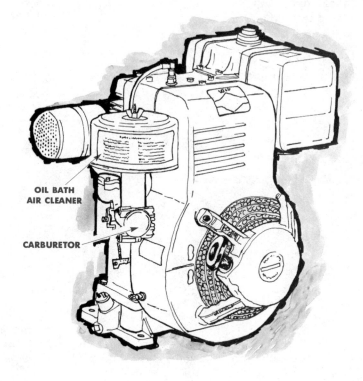

Figure 22-1: Oil-bath air cleaners are mounted on top of carburetors with a wing nut.

OIL BATH
AIR CLEANER

CARBURETOR

when the air goes through an oiled screen at the top of the air cleaner can. Oil-bath air cleaners cannot be turned on their side because the oil would run out.

The <u>dry element air cleaner</u> (figure 22-3) will work in any position. It is used on many small engines. Air en-

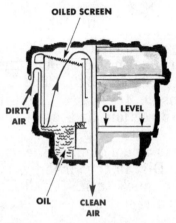

Figure 22-2: In an oil-bath air cleaner, the sharp change in air flow direction just above the oil-bath causes dirt to fall into the oil. The air is finally filtered through an oiled screen.

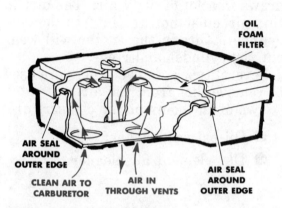

Figure 22-4: Oil-foam filters are like a sponge. Air can pass through the tiny holes, but dirt is trapped in the filter.

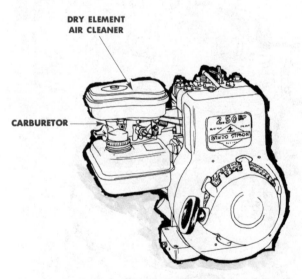

Figure 22-3: Dry element air cleaners attach to the top of a carburetor with screws.

Figure 22-5: Paper filters are used in some dry air cleaners. Air passes through the holes in the paper while dirt is trapped outside.

ters the dry air cleaner through large holes in the bottom of the can. Then it goes through the filter.

There are three types of dry element air cleaners. Each has a different type of filter.

The oil-foam filter is like a sponge. Air can pass through the small holes but the dirt is trapped (figure 22-4).

Some dry element air cleaners use a paper filter (figure 22-5). The paper filter has tiny holes in it. Air can get through the filter paper, but dirt sticks to the outside.

Another type of dry element air cleaner uses a filter made of aluminum foil (figure 22-6). Strips of lightly oiled aluminum foil are pressed together. Air must make its way through the strips. When it does, dirt sticks to the oil.

SELF CHECK

1. What will dirt do to engine parts?
2. What does the air cleaner do?
3. How does an oil-bath air cleaner work?
4. Why do some engines use a dry air cleaner?

Activities for Unit 22

1. Take an oil-bath air cleaner apart and explain how air gets through.

2. Take apart an oil-foam air cleaner and explain how air gets through.

3. Take apart a paper filter air cleaner and explain how air gets through.

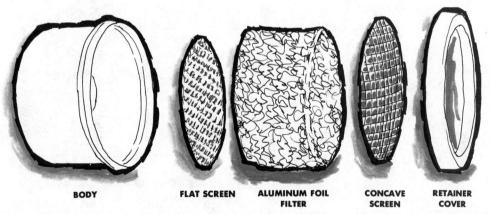

BODY   FLAT SCREEN   ALUMINUM FOIL FILTER   CONCAVE SCREEN   RETAINER COVER

Figure 22-6: Aluminum foil air cleaners use oiled aluminum foil filters. The dirt passing through the filter sticks in the oil.

An engine must have just the right amount of fuel to start and run well. Too much or too little fuel will cause the engine to run roughly or not run at all.

As noted in Unit 14, the first thing to check when an engine is running roughly, or not at all, is the spark. The second thing is the fuel. Working with fuel usually means working with gasoline. **Gasoline can explode. Do not work near open flames or fires. Do not let anyone smoke near gasoline. Always open doors or windows in the work area. Do not breath the vapors.**

The first step in finding a fuel problem is to take out the spark plug with a spark plug wrench (figure 23-1). Examine the spark plug. It should be brown or tan in color and look fairly

Figure 23-2: The right air-fuel mixture leaves spark plugs brown or tan in color. (Champion)

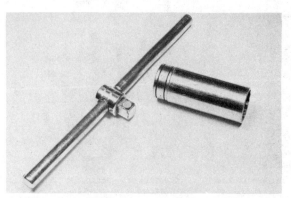

Figure 23-1: Some fuel problems can be found and solved with a spark plug wrench.

Figure 23-3: A black, wet spark plug means the cylinder is getting too much fuel. A spark plug wet with gasoline will not spark. (Champion)

clean (figure 23-2). If the color is wrong or the spark plug is wet, there may be a fuel problem.

An engine with <u>too much fuel</u> in the cylinder will not run right. It will run rough. When the spark plug is wet with unburned gasoline it will not spark (figure 23-3).

A common cause of too much fuel is overchoking. Move the choke lever to make sure it isn't stuck in the closed position and is properly set (figure 23-4).

Figure 23-5: When the spark plug is covered by a dry, white ash, it is not getting enough fuel. (Champion)

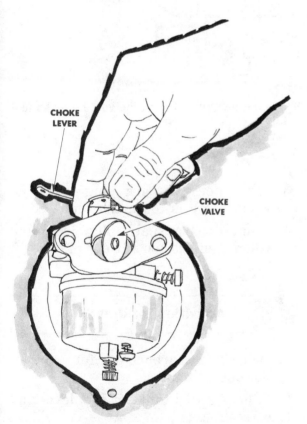

Figure 23-4: Too much fuel can be caused by the choke. It can be set in the wrong position or it may stick in the wrong position.

An engine that will not start or run <u>may not be getting enough fuel.</u> A spark plug from an engine that is not getting enough fuel will be dry and white (figure 23-5).

Take the cap off the fuel tank. There should be <u>fuel in the tank.</u> Smell the gasoline. When it sits too long it gets stale and smells like paint. Stale gasoline can plug up the carburetor.

If there is a <u>fuel tank valve,</u> make sure it is open (figure 23-6).

<u>Dirt or grass in the fuel tank</u> can plug the fuel line at the bottom of the tank. The screen that prevents dirt from entering the carburetor may be covered with dirt. A <u>plugged fuel cap vent</u> will stop fuel from running to the carburetor by creating a vacuum in the fuel tank.

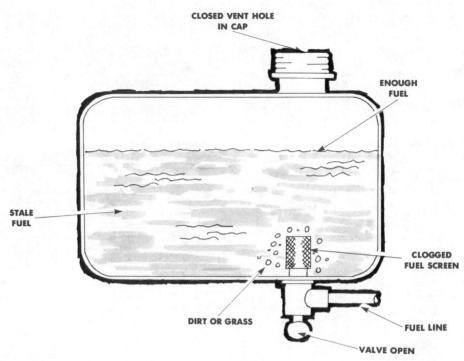

CLOSED VENT HOLE
IN CAP

ENOUGH
FUEL

STALE
FUEL

CLOGGED
FUEL SCREEN

DIRT OR GRASS

FUEL LINE

VALVE OPEN

Figure 23-6: Always check to make sure there is fuel in the fuel tank. Check the vent hole and fuel screen for clogging. Smell the gasoline to see if it is stale.

The carburetor high and low speed adjustment screws may be improperly set. They can cause too much fuel in the engine, or not enough.

3. What does a spark plug look like when there is too much fuel?

4. What does a spark plug look like when there is too little fuel?

Activities for Unit 23

1. Check a carburetor for correct choke operation.

2. Check a fuel tank for fuel level and fuel freshness.

3. Remove a fuel line and check for a plugged tank filter.

SELF CHECK

1. Give two reasons why an engine will run roughly.

2. What does a spark plug look like when the air-fuel mixture is right?

The air cleaner traps dirt before it gets into the engine. But the dirt plugs up the air cleaner. Dirty air cleaners may keep the engine from getting enough air. With too little air, an engine may not start or run well.

Air cleaners are usually mounted on the carburetor. Different kinds of air cleaners are mounted differently. They are cleaned with common tools (figure 24-1). The three most common kinds of air cleaners are:

- Oil-foam
- Oil-bath
- Dry paper

Most oil-foam air cleaners are held on by a screw. The screw goes through the air cleaner and into the carburetor.

Figure 24-1: Air cleaners are cleaned with the simplest elements: a screwdriver, a bucket of soap and water, and oil.

Figure 24-2: Some air cleaners are held on with screws or have wing nuts which can be undone by hand.

Figure 24-3: Oil-bath and many dry-element air cleaners are held on by wing nuts.

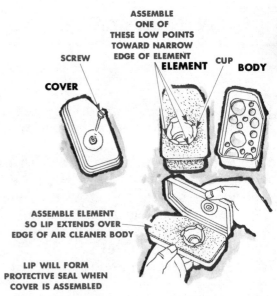

ASSEMBLE
ONE OF
THESE LOW POINTS
TOWARD NARROW
SCREW   EDGE OF ELEMENT   CUP   BODY
ELEMENT

COVER

ASSEMBLE ELEMENT
SO LIP EXTENDS OVER
EDGE OF AIR CLEANER BODY

LIP WILL FORM
PROTECTIVE SEAL WHEN
COVER IS ASSEMBLED

Figure 24-5: Oil-foam air cleaners are taken apart by lifting off the cover.

Figure 24-4: Care should be taken not to drop dirt in the carburetor when removing the air cleaner.

Use a large screwdriver to take out the screw (figure 24-2). Then lift the air cleaner off the carburetor. Avoid dropping dirt into the carburetor.

Oil-bath air cleaners and many dry paper air cleaners are held on by a wing nut. The nut fits on a shaft that sticks up from the carburetor (figure 24-3). Take off the wing nut. Lift the air cleaner off the carburetor without spilling any dirt (figure 24-4).

The oil-foam air cleaner has three parts: the cover, the element, and the body (figure 24-5). Lift off the cover. Take the oil-foam element out of the body. Wash the cover and body in a bucket of soapy water. Dry the parts with a rag.

Wash the foam in soapy water.

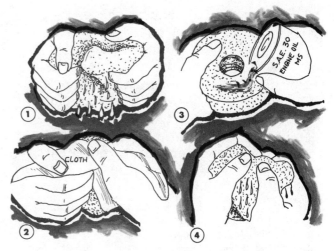

Figure 24-6: Oil-foam air cleaner elements are cleaned by (1) washing them in soapy water, (2) squeezing them dry with a cloth, (3) filling them with clean oil, and (4) squeezing them dry.

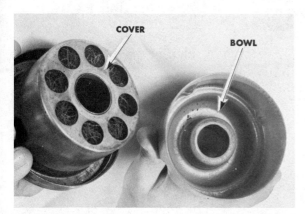

Figure 24-7: In this type of air cleaner, the oil is in the oil-bath bowl.

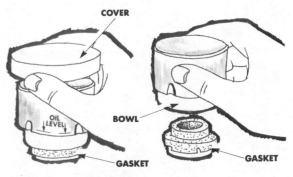

Figure 24-8: This oil-bath air cleaner has a cover, bowl and gasket. The oil screen is in the cover.

Rinse it in clean water. Wrap it in a dry rag and squeeze it dry. Pour engine oil on the foam. Then squeeze the foam to get out most of the oil (figure 24-6).

The oil-bath air cleaner has two parts, a cover and a bowl (figure 24-7). First lift off the cover (figure 24-8).

You will see that the bowl is filled with dirty oil. Dump this oil, and wipe the bowl clean with a rag. Refill the bowl with clean oil. There is a line on the bowl marked "oil level". Fill the bowl to this line. Do not overfill the bowl or the oil will be drawn into the carburetor. Clean the cover and put it back on the bowl.

Dry paper air cleaners have two parts: a cover and a paper filter element (figure 24-9). Remove the screw, and lift the cover off. Tapping the paper element on the bench will knock off a lot of dirt. Wash the filter element in soapy water and then pour clean water through the filter from the inside. Dry the filter before you put it back.

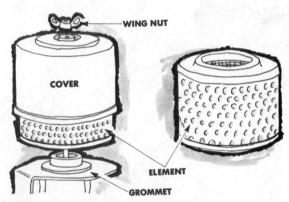

Figure 24-9: The dry paper air cleaner is taken apart by unscrewing the wing nut and lifting off the cover and the paper filter.

SELF CHECK

1. What happens when too much dirt builds up in an air cleaner?
2. What is used to clean air cleaner parts?
3. Why must an air cleaner be taken off carefully?
4. How is a dry paper air cleaner cleaned?

Activities for Unit 24

1. Remove, clean and replace an oil-bath air cleaner.

2. Remove, clean and replace an oil-foam air cleaner.

3. Remove, clean and replace a paper air filter.

ADJUSTING A CARBURETOR

Unit 25

An engine needs just the right amount of air-fuel mixture. Too much or too little causes the engine to run rough. From time to time the air-fuel mixture must be adjusted at the carburetor.

Most carburetor adjustment is done with three screws (figure 25-1). They are:

- High speed adjustment screw
- Low speed adjustment screw
- Idle speed screw

The vacuum carburetor has two adjusting screws (figure 25-2). The high speed adjustment controls the fuel coming up the fuel pipe. The idle speed

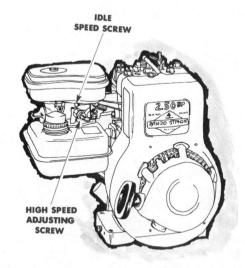

Figure 25-2: The vacuum carburetor has two adjusting screws: the high speed adjustment screw and the idle speed screw.

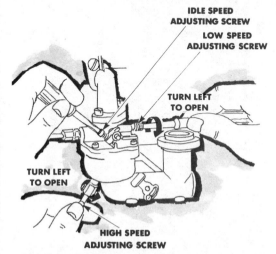

Figure 25-1: Most carburetor adjustments can be made with a screwdriver.

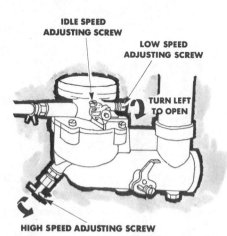

Figure 25-3: On this float carburetor the low speed adjustment screw and the high speed adjustment screw are clearly visible.

93

screw controls how far the throttle valve can close, setting the engine's idle speed.

The float type carburetor has three adjusting screws (figure 25-3). It has a high speed and a low speed adjusting screw. It also has an idle speed adjusting screw.

A diaphragm type carburetor (figure 25-4) has the same three adjusting screws as the float carburetor. Most carburetors are adjusted in the same way.

To set the high speed adjustment screw, start the engine. Warm it up for several minutes. Find the carburetor high speed adjusting screw. Some screws have a screwdriver slot. Others can be adjusted by hand.

Open the engine's throttle until it runs at high speed. Turn the high speed adjustment screw in to the right (figure 25-5). This closes off the flow of fuel to the venturi. Slowly turn the screw until the engine runs rough. Hold the throttle steady. Turn the adjusting screw out slowly to the left. The engine will speed up. Turn the adjusting screw too far and the engine will begin to slow down. Stop turning when the engine runs as fast as it will run.

To set the low speed adjustment screw, start with a warm engine. Close the throttle and allow the engine to idle.

Find the low speed adjusting screw. Put the screwdriver into the low speed adjusting screw. Slowly turn the screw

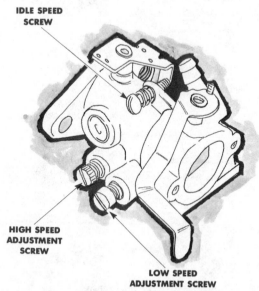

Figure 25-4: The three basic carburetor adjustment screws are shown on this diaphragm carburetor: idle speed screw, high speed adjustment screw and low speed adjustment screw.

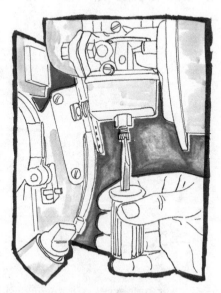

Figure 25-5: Turning the high speed adjustment screw controls the amount of fuel allowed to pass through the fuel pick-up into the carburetor.

Figure 25-6: The low speed adjustment screw controls the fuel being delivered to the carburetor behind the throttle valve.

Figure 25-7: The idle speed adjustment screw works differently than other carburetor adjustments. It physically stops the throttle from closing beyond a preset point.

in (to the right) (figure 25-6). This shuts off the low speed fuel. Stop turning when the engine runs rough. Turn the screw slowly out (to the left). This lets more mixture into the engine. The engine will begin to speed up. Stop turning the screw when the engine runs as fast as it will. If you go too far the engine will slow down.

To set the idle speed adjustment start with a warm engine. Find the idle speed screw. With a screwdriver turn the screw to the right (figure 25-7). This opens the throttle valve. The engine should just run fast enough so that it will not stop. To slow it down, turn the screw to the left.

SELF CHECK

1. Name the adjustment screws of a vacuum carburetor.
2. Name the adjustment screws of a float carburetor.
3. Name the adjustment screws of a diaphragm carburetor.
4. How is the high speed adjustment screw adjusted?

Activities for Unit 25

1. Adjust the idle speed of a carburetor.
2. Adjust the low speed of a carburetor.
3. Adjust the high speed of a carburetor.

When an engine runs it develops a great deal of heat. All engines have a cooling system to remove the heat. All engines must also have a lubrication system to get oil between their operating parts to help them last longer.

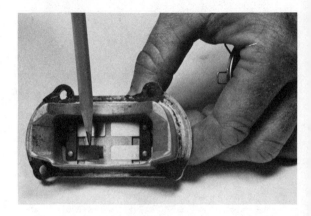

Figure 26-1: Large engines, like automobile engines, are water-cooled. Liquid cooling is more efficient than air cooling, but it increases the engine weight. (Chrysler)

During the engine's power stroke a mixture of air and fuel is burned in the cylinder. This burning causes a lot of heat. If the heat is not taken away, the engine will be damaged.

There are two ways to cool an engine:

- Liquid cooling
- Air cooling

In a <u>liquid cooled</u> engine, water is forced through passageways in the engine block and over the hot engine parts. The water takes away the heat.

Figure 26-3: Motorcycles depend on their own movement through the air to cool their engines. (BMW)

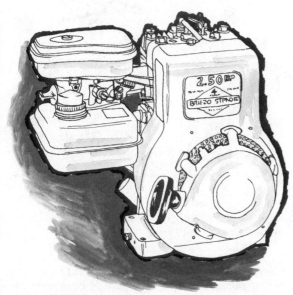

Figure 26-2: Most small engines are air-cooled.

Most automobile engines are water-cooled (figure 26-1).

Most small engines are air-cooled (figure 26-2). The heat is removed by air that is forced around the hot engine parts in two ways. The engine may be forced through the air, like a motorcycle engine (figure 26-3). Or, the air may be forced over the engine like a lawn mower engine (figure 26-4). An air pump or fan is used to force air over these engines.

Cooling fins and flywheel fins (figure 26-5) are the main elements of a small engine cooling system. They must be kept free of heavy grease and dirt. Dirt and grease on the fins prevent the heat from flowing into the air. Failure to

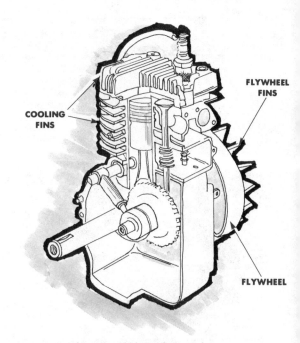

Figure 26-5: Basic cooling aids on small air-cooled engines are the fins on the cylinder head, cylinder, and flywheel.

Figure 26-4: Most engines require some method of moving air over their cylinders to control heating problems. This tractor does not move through the air fast enough to cool its engine. A fan must be used. (Toro)

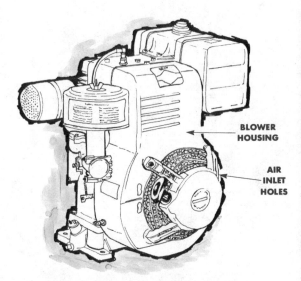

Figure 26-6: The flywheel fins operate as an air fan or pump. When the engine is operating, the fins direct a stream of moving air over hot parts of the engine.

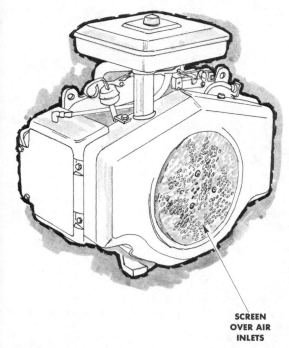

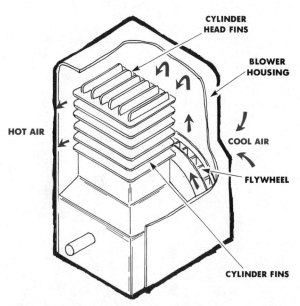

Figure 26-9: This is how the air flows from the flywheel through the cooling fins. If the air didn't move across the engine surfaces, they would get hot and create a heating problem.

Figure 26-7: In order to blow air out, the engine must take it in. A screen over the air inlet keeps solid debris from being drawn into the flywheel fins.

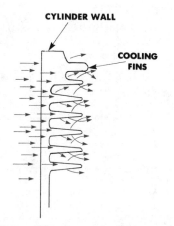

Figure 26-8: This diagram shows how heat moves from the cylinder to the cooling fins of air-cooled engines.

keep them clean can result in damage to the engine due to overheating.

The running engine turns the flywheel. The fins on the flywheel work like a fan (figure 26-9). Cool air is pulled into the inlet hole of the blower housing and is blown over the engine. The heat is transferred from the cooling fins to the circulating air and exhausted away from the engine.

The shape of the blower housing controls the flow of air. In figure 26-10, for example, air is forced to the top of the blower housing and through each cooling fin. This removes heat from the fins.

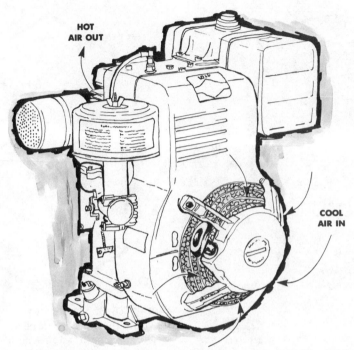

HOT AIR OUT

COOL AIR IN

Figure 26-10: This diagram shows how cool air is drawn into the air inlet and exhausted on the other side.

SELF CHECK

1. What are the two ways of cooling small engines?

2. How does liquid cooling work?

3. Why must cooling fins be kept clean?

4. How does the flywheel help air cooling?

Activities for Unit 26

1. List all the engines you can think of that use liquid cooling.

2. List all the engines you can think of that use air cooling.

3. Using a cut-away engine model, show how air passes over an engine to cool it.

Many small engines are run in places where there is a lot of dirt or grass in the air. Power lawn mowers and chainsaws are good examples. Dirt and grass in the air are pulled into the blower housing. They stick to the cooling fins, and trap engine heat. This trapped heat may damage the engine.

Do not touch an engine that has been running. It can be hot enough to cause a bad burn. Let the engine cool before you clean the cooling fins.

When the engine is cool, remove the blower housing. The blower housing is usually held to the engine with three bolts or screws (figure 27-1). Use an open-end wrench or a screwdriver to remove them. One bolt or screw is at the top of the blower housing. The

Figure 27-2: Plugged cooling fins are easy to find. Just take a look at them.

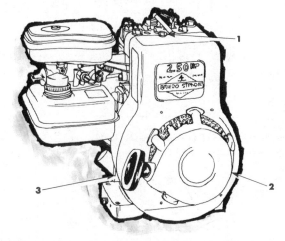

Figure 27-1: Blower housings are generally held on by three bolts. Sometimes the housing may be held on with screws.

other two are along the side. They may be hard to find if they are covered with grass and dirt. Turn the wrench or screwdriver to the left to remove the bolts (figure 27-2).

Lift off the blower housing. Tap the blower housing on the bench. This will shake some of the grass and dirt free.

Examine the cooling fins (figure 27-3). Are they plugged with grass or dirt like those shown in figure 27-2? If they are, scrape them clean with a scraper (figure 27-4).

Now is a good time to check the governor air vane. Move it back and

Figure 27-3: Use a scraper to remove grass and dirt from plugged cooling fins.

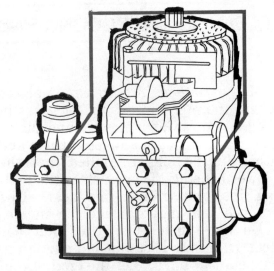

KEEP AREAS WITHIN HEAVY LINE CLEAR OF ALL GRASS AND DIRT

forth with your hand to make sure it is free.

When all the dirt and grass is off the fins and blower housing, the housing may be put back on. Start the three bolts or screws by hand. Tighten them with a wrench or screwdriver.

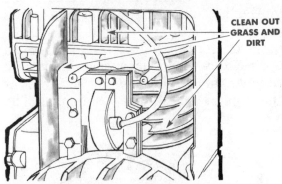

CLEAN OUT GRASS AND DIRT

Figure 27-4: This drawing shows the areas that must be kept clean for good cooling.

SELF CHECK

1. What happens if the cooling fins are plugged?

2. Why should you allow the engine to cool before you work on it?

3. Where are the blower housing bolts or screws usually found?

4. How is the governor air vane checked?

Activities for Unit 27

1. Remove a blower housing from a small engine.

2. Clean the cooling fins of a small engine.

3. Replace the blower housing of a small engine.

Oil keeps the moving parts of an engine from rubbing on one another. The parts do not rub together, they are separated by a very thin film of oil. They move easier because oil is very slick and reduces friction (figure 28-1). Reducing friction helps to reduce heat and wear.

The oil also helps cool the engine by carrying heat away from the hot engine parts. Oil on the cylinder walls helps the rings stop compression leaks. Oil also reduces friction, and helps to cool the engine.

Getting the oil to the engine parts is called lubrication. One kind of lubrica-

tion system used in small engines is called splash lubrication.

Two types of splash lubrication systems are used in small engines:

- Dipper systems
- Slinger systems

Horizontal crankshaft engines use an <u>oil dipper</u>. The dipper hooks to the bottom of the connecting rod (figure 28-2). When the rod goes around, the dipper goes around with it. It dips into the oil in the crankcase and splashes oil on the engine parts.

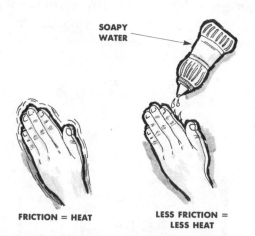

Figure 28-1: By rubbing your hands together you can generate friction and heat. It you put a slippery liquid between your hands, there will be less heat and your hands will move faster. This is because your hands actually press against a film of lubricant, not against each other. The lubricant is more slippery than your hands.

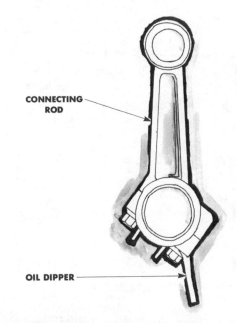

Figure 28-2: Horizontal crankshaft engines use a dipper on the connecting rod to splash oil on their parts.

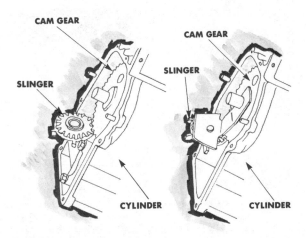

Figure 28-3: Vertical crankshaft engines lubricate with an oil slinger. It is driven by the camshaft.

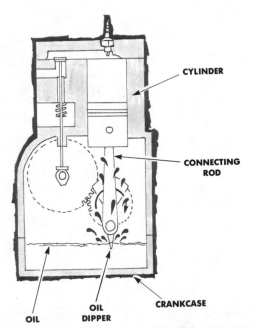

Figure 28-4: Dippers and slingers both operate by splashing oil up inside the crankcase.

Vertical crankshaft engines cannot use a dipper. The reason for this is that the connecting rod is not near the oil in the crankcase. A small gear called a slinger is used to splash oil on the engine parts. The slinger is a gear that rides on the camshaft gear. When the camshaft turns, the slinger turns (figure 28-3). The small paddles on the slinger dip into the oil and splash it on the engine parts.

Both the vertical and horizontal engines' crankcases are half filled with oil. As the engine runs, oil is splashed upward by the dipper or slinger (figure 28-5).

Some oil is splashed on the cylinder walls. It is carried up the wall by the piston and rings. This provides lubrication between the piston and the cylinder wall.

The oil control rings act like a squeegee. They wipe the cylinder wall when the piston is traveling down. Oil passes through the spaces in the oil

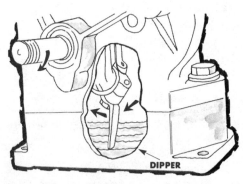

Figure 28-5: As the crankshaft turns, the dipper splashes through the oil at the bottom of the crankcase and throws it on the engine parts.

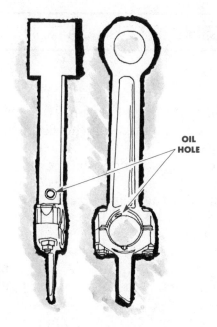

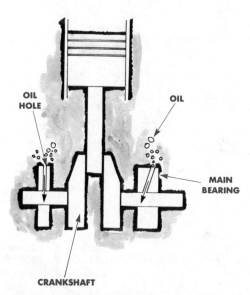

Figure 28-6: Oil holes in engine parts let oil get inside for lubrication. As oil runs back to the crankcase it flows into the oil holes. Here the oil holes are used to lubricate the main bearings.

control ring, into the holes in the piston, and back into the crankcase.

The splashed oil also runs down to the crankcase through special holes in some of the engine parts. Oil running into these holes lubricates the main bearings, rod bearing, and crankshaft (figure 28-6).

SELF CHECK

1. List three reasons why oil is needed between engine parts.
2. What do the oil control rings do?
3. What is an oil dipper?
4. How does oil get to the cylinder walls?

Activities for Unit 28

1. Using a cut-away engine model, trace the flow of oil in a horizontal crankshaft engine.

2. Using a cut-away engine model, trace the flow of oil in a vertical crankshaft engine.

3. Put a drop of oil between two pieces of metal and rub them together. What happens to friction?

In pressure lubrication a pump is used to force oil into engine parts. It works better than splash lubrication because the oil is <u>forced</u> into the engine parts. Most small engines use a barrel oil pump (figure 29-1). Barrel oil pumps have two parts:

● The body

● The plunger

The <u>body</u> is the larger part of the pump. The <u>plunger</u> fits inside the body.

The pump operation can be compared to a piston and cylinder. When the plunger moves out of the body it creates a vacuum. This is called the intake stroke. When the plunger moves into the body, it is called the compression stroke. The pump pulls in oil on the intake stroke. It pushes oil into engine parts on the compression stroke.

The pump plunger has a large off-center hole in it. This off-center hole fits around the engine's camshaft. The body fits on a shaft that does not move (figure 29-2).

When the engine is running the camshaft goes around. A cam lobe on the camshaft causes the plunger to

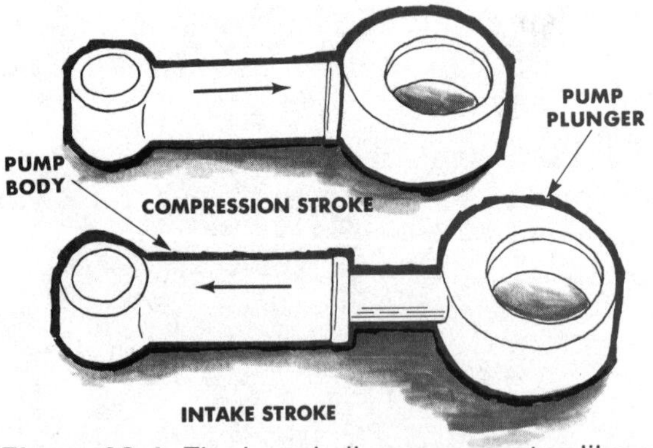

PUMP PLUNGER

PUMP BODY

COMPRESSION STROKE

INTAKE STROKE

Figure 29-1: The barrel oil pump operates like a piston and cylinder. It takes in oil on the intake stroke and squirts it out on the compression stroke.

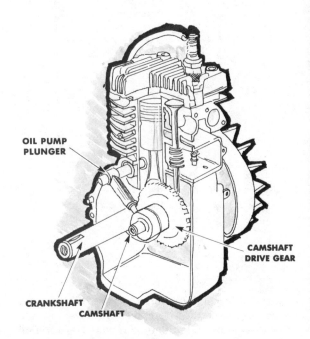

OIL PUMP PLUNGER

CAMSHAFT DRIVE GEAR

CRANKSHAFT

CAMSHAFT

Figure 29-2: Note how the oil pump plunger fits over the camshaft. As the camshaft turns, the plunger is forced in and out of the body by a cam lobe on the camshaft.

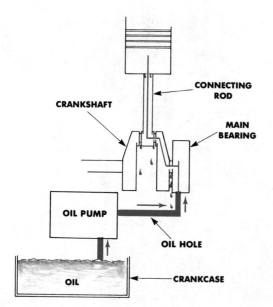

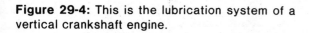

Figure 29-3: The path of lubricating oil through the oil holes in engine parts can be very complicated.

move up and down. The body cannot move because it is attached to the crankcase. The plunger slides in and out of the pump body. Oil is pulled into the pump body on the intake stroke. On the compression stroke, oil is forced out of the pump into the engine parts.

The pump pulls oil out of the crankcase on the intake stroke. On the compression stroke oil is pushed out of the pump through an oil hole (figure 29-3).

The oil then passes through another hole into the main bearings. This gives the main bearings their lubrication. Another hole in the crankshaft lets oil flow from the main bearing to the connecting rod bearing.

Some engines have an oil hole that goes up the center of the connecting rod to lubricate the piston pins.

Figure 29-4: This is the lubrication system of a vertical crankshaft engine.

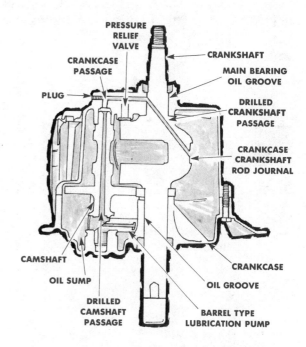

Figure 29-5: Car engines are so complex that they require pressure lubrication.

When the engine is running, oil passes between the crankshaft and connecting rod. The moving crankshaft throws this oil up onto the cylinder wall.

SELF CHECK

1. What makes oil flow in a pressure lubrication system?
2. What kind of oil pump is used in most small engines?

3. Name the two parts of an oil pump.
4. Which engine part makes the pump work?

Activities for Unit 29

1. Take apart a barrel oil pump and name the parts.

2. Using a cut-away engine model, show how an oil pump works.

3. Using a cut-away engine model, trace the oil flow of a small engine pressure lubrication system.

Most two-stroke cycle engines do not use splash or pressure lubrication. There are two methods of lubricating a two-stroke cycle engine:

● Fuel mixture

● Oil injection

Many small two-stroke cycle engines use a mixture of fuel and oil (figure 30-1). Usually they are mixed together in the proper amounts before the mixture is poured into the fuel tank.

The <u>fuel mixture</u> enters the engine's crankcase through the reed valve (figure 30-2). It enters the crankcase as a spray. Some of this spray falls on the engine parts. The oil that lubricates the engine is in the mixture.

Some of the oil goes into the combustion chamber with the fuel, and is burned.

<u>Oil injection</u> is another way to get the fuel and oil mixed together. Larger two-stroke cycle engines have two tanks. The fuel tank holds the fuel. Another tank holds the oil.

When the reed valve is open, air and fuel enter the crankcase. A pump forces a small amount of oil into the crankcase at the same time (figure 30-3). The oil enters the crankcase through a passageway very close to the reed valve. The oil, fuel, and air mix in the crankcase.

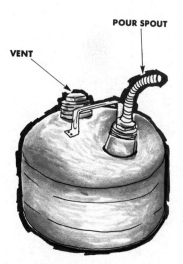

Figure 30-1: Fuel and oil are mixed together (and stored) in a clean, properly marked can with a vented cap and a pouring spout.

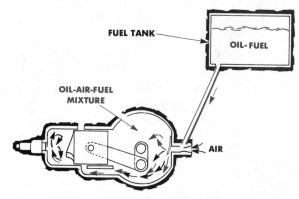

Figure 30-2: A mixture of oil and fuel goes into the crankcase with the air. Some oil is carried into the combustion chamber where it is burned.

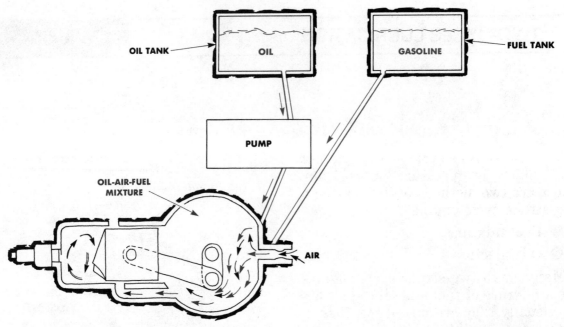

Figure 30-3 With fuel injection, the oil is injected into the crankcase from a separate oil tank.

Two-stroke cycle oil is a special type of oil. It mixes with gasoline and stays mixed for a long time. The mixture is different for different engines. Some engines use 20 parts of gasoline for each part of oil. Other engines use 50 to 1 mixtures. This means 50 parts of gasoline are used with one part of oil. There is always more gasoline than oil in the mixture. Premix is fuel which is sold with the oil already mixed in it.

SELF CHECK

1. Name two ways two-stroke cycle engines are lubricated.

2. What does a "50 to 1 mixture" mean?

3. How does oil get on two-stroke cycle engine parts?

4. What is oil injection?

Activities for Unit 30

1. Look up the proper mixture of fuel and oil for a two cycle engine.

2. Mix fuel and oil together for that two cycle engine.

3. Using a cut-away engine model, trace the flow of the fuel mixture in a two cycle engine.

An engine won't run very long without oil. If the oil in an engine is dirty it can scratch the internal parts and ruin the engine. So it is important to check and change the oil often.

To <u>check the oil level</u> you must find the oil filler plug. It is one of two plugs on the outside of the engine (figure 31-1). The higher plug is the oil filler plug. The oil filler plug on most small engines is made of plastic. Sometimes it is shaped like a wing nut.

To check the oil, put the engine on a level surface. Remove the oil filler plug by turning it to the left. Look in the oil filler plug hole. The oil should be up to the top of the filler hole. If it isn't, add oil. Start by putting the small end of a funnel in the filler hole. Slowly pour the oil into the large end of the funnel (figure 31-2). Stop pouring when the oil starts filling the funnel. The oil should only come up to the top of the filler hole.

To <u>change dirty oil</u>, start the engine and allow it to warm up. Move the engine over a drain pan so that the dirty oil will run into the drain pan, not

Figure 31-1: Engine crankcases have a filler plug and a drain plug.

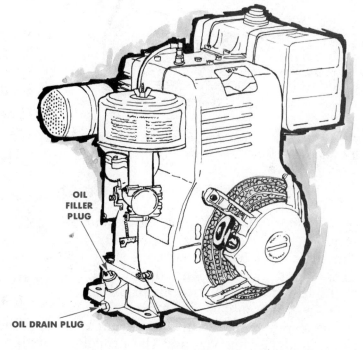

OIL
FILLER
PLUG

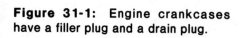
OIL DRAIN PLUG

Figure 31-2: Fill the crankcase through the filler oil plug hole.

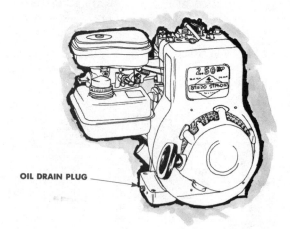

Figure 31-3: The drain plug is on the bottom of the crankcase.

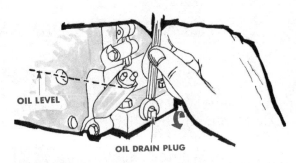

Figure 31-4: To remove the oil, remove the drain plug with a wrench.

kind of oil. Most small engines use SAE 30 (weight) oil, but some are different. You should check. The wrong oil can harm an engine.

SELF CHECK

1. Why is dirty oil in a small engine dangerous?
2. How do you check the oil level of a small engine?
3. How is an oil drain pan used?
4. Where do you fill the crankcase with oil?

on the floor, when the plug comes out. Remove the drain plug (figure 31-4). Tip the engine to make sure all the oil drains out. Replace the drain plug. Refill the crankcase with clean oil the same way you added oil before.

Check the manufacturer's specifications to make sure you add the proper

Activities for Unit 31

1. Remove the oil drain plug from a small engine and drain the oil.
2. Refill a small engine with oil.
3. Check the oil level in a small engine.

At least one out of every five American workers has a job in the field of power mechanics. That means they work in one of the three following areas:

- Power mechanics production
- Power mechanics sales
- Power mechanics service

Production is also called manufacturing. Production is the process of changing raw materials such as steel, iron, plastic, and glass into a finished product. Production workers in the power mechanics field make airplanes, cars, motorcycles, boats, and many other mechanical items.

Sales is the second major power mechanics career area. Salespersons demonstrate and explain the products that are manufactured by the production workers.

Service of power mechanics products means repairing them. Service people are usually called mechanics. They fix or adjust almost any product with an engine.

Preparing for a power mechanics career starts with doing a good job in school. Mechanics must be able to read well so they can follow directions.

Reading about jobs, and talking to people who have them, is the best way to learn about careers. Parents, teachers, and your school counselors are good sources of information.

Different careers require different skills and interest. School is a place to learn skills. But only you can provide the interest.

SELF CHECK

1. What is production?
2. List three careers in power mechanics.
3. What does a mechanic do?
4. How would you find out about a career?

Activities for Unit 32

1. Visit a place where small engines are sold. Report on what you see.

2. Visit a place where small engines are manufactured. Report on what you see.

3. Visit a place where small engines are repaired. Report on what you see.

GLOSSARY

Air cleaner: A filter mounted above the carburetor. Used to clean air before it enters the engine.

Air cooling: Cooling engine parts by circulating air around them.

Bearing: A part in which another part turns. Used to reduce friction and wear between moving parts.

Block: A metal box containing the crankcase and cylinders of an engine.

Bolt-on cylinders: Cylinders that are held in place with bolts or studs. This type of cylinder can be removed for service.

Bore: The diameter of the cylinder.

Breaker points: The switch used to control the operation of the coil.

Bushing: The type of bearing used on the piston end of the connecting rod.

Camshaft: A shaft with lobes that are used to open an engine's valves at the proper time.

Carburetor: A part that mixes and meters air and fuel in the correct amounts to burn in the engine.

Coil: An electrical device used to step-up voltage. The coil of a small engine provides the spark to ignite the air-fuel mixture.

Combustion chamber: The part of the engine in which the burning of air and fuel takes place.

Compression ring: The piston ring used to seal the compression pressures in the combustion chamber.

Compression stroke: The stroke of the four-stroke cycle in which the air-fuel mixture is compressed.

Condenser: A magneto part used to prevent contact point arcing by storing electrons.

Connecting rod: An engine part that connects the piston to the crankshaft.

Connecting rod bearing: The bearing used between the connecting rod and the crankshaft.

Coolant: The liquid used in a liquid cooling system to carry away heat.

Cooling fins: The metal fins on air cooled engine parts. Used to transfer heat away from the parts.

Crankcase: The part of the engine that holds the crankshaft.

Crankshaft: An offset shaft to which the piston and connecting rods are attached.

Cylinder: The hole in which the engine's pistons travel.

Cylinder head: The part bolted to the top of the engine. It forms the top of the combustion chamber.

Diaphragm carburetor: A carburetor that uses a diaphragm pump to control the fuel supply to the engine.

Displacement: The inside area of an engine (Bore times Stroke).

Engine: Any machine that burns fuel to make power.

Exhaust ports: The passages used to carry burned gases out of the cylinder.

Exhaust stroke: The stroke of a four-stroke cycle engine in which the burned gases are pushed out of the cylinder.

Exhaust valve: The valve used to allow the flow of burned exhaust gases out of the cylinder.

External combustion engine: An engine in which the burning of fuel takes place outside the engine.

Float: The part of the carburetor that provides the correct amount of fuel to the float bowl.

Gear: A wheel with teeth that are used to transfer rotating motion to another gear.

Governor: A device used to control engine speed and load.

Ground electrode: The spark plug electrode that is connected to ground.

High speed adjustment: The screw on a carburetor used to adjust fuel mixture at high speed.

Horsepower: A term used to describe the power developed by engines. One horsepower is equal to 33,000 foot-pounds of work per minute or 550 foot-pounds per second.

Intake ports: The passages used to route the flow of air and fuel into the cylinder.

Intake stroke: The stroke of the four-stroke cycle engine in which the air-fuel mixture enters an engine.

Intake valve: A valve used to control the flow of air and fuel into the engine.

Internal combustion engine: An engine in which the burning of fuel takes place inside the engine.

Journal: The part of a shaft around which a bearing fits.

Lobe: A raised bump on a cam shaft. Used to lift valves.

Low speed adjustment: The screw on a carburetor used to adjust mixture at low speed.

Lubrication: Reducing friction in an engine by providing an oil film between moving parts.

Magneto: The device used to make the spark which ignites the air-fuel mixture of an engine.

Oil: A petroleum-based liquid used to provide lubrication.

Oil control ring: A piston ring used to keep oil from getting into the combustion chamber.

Oil filter: The device used to filter out dirt and other foreign matter from the oil.

Oil pump: The device used to circulate oil to the moving parts of an engine.

Piston: The round metal part which slides up and down in the cylinder.

Piston pin: The pin that attaches the piston to the connecting rod.

Piston ring: The sealing rings placed in grooves around the piston.

Power stroke: The stroke of the four-stroke cycle engine in which power is delivered to the crankshaft.

Retainer: A washer and lock assembly used to hold the valve spring in position.

Spark plug: A part used to create the spark in the combustion chamber.

Splash lubrication: The process of lubricating engine parts by splashing oil on them.

Stroke: The travel of the piston in the cylinder, controlled and measured by the offset of the crankshaft.

Vacuum carburetor: A carburetor that uses a vacuum to draw fuel out of the fuel tank.

Valve: A device used to open and close ports.

Valve guide: The part which is installed in the cylinder block to support and guide the valve.

Valve lifter: The device that rides on the cam and pushes on the valve.

Valve seat: The part of the cylinder head that the valve seals against.

Valve spring: The coil spring used to close the valve.

Venturi: A restricted (constricted) area in the carburetor used to develop the vacuum that draws in fuel.

Wrist pin: See Piston pin.

INDEX